MOON
HANDBOOK

A 21ST-CENTURY TRAVEL GUIDE

MOON HANDBOOK

A 21ST-CENTURY TRAVEL GUIDE

CARL KOPPESCHAAR

**TRANSLATED BY
SUSAN MASSOTTY**

MOON HANDBOOK, A 21ST-CENTURY TRAVEL GUIDE

Published by
Moon Publications, Inc.
P.O. Box 3040
Chico, California 95927-3040, USA

Printed by
Colorcraft Ltd., Hong Kong

© Copyright Uitgeverij J.H. Gottmer/H.J.W. Becht bv, Haarlem, The Netherlands, 1993.
Originally published under the title: *De Maan (Dominicus reisgids)*.

© Translation copyright Susan Massotty, 1995.

First English edition, 1995.
All rights reserved.

Text: Carl Koppeschaar
Cartography: Wil Tirion/De Koepel Foundation (Utrecht, Netherlands), pp. 14–15, 68; Carl Koppeschaar, pp. 21, 38, 73, 96; Bob Race, p. 95

Photos and illustrations: Boeing Defense & Space Group, p. 103; Carl Koppeschaar, color photo facing p. 81 (upper); black-and-white illustrations pp. 18, 24, 42, 43, 45, 47, 49, 50, 56, 75, 80, 90; Don Dixon, color illustration facing p. 81 (lower); Erven J.J. Tijl Press (Zwolle, Netherlands), color illustrations facing pp. 48 (upper), 64 (upper and lower), black-and-white illustrations pp. 98, 102; Hergé Foundation/Casterman Press, pp. 12, 81; Jinsei Choh/Ohbayashi Corporation, color illustration facing p.65 (lower); Lee Battaglia, color illustration facing p. 49; Mount Wilson Observatory, pp. 28, 29; NASA/Lyndon B. Johnson Space Center, pp. 36, 94, 105, 107, 109; NASA, color photo facing p. 48 (lower); black-and-white photos pp. 33, 59, 62, 86, 115; Pat Rawlings/NASA, color illustrations facing pp. 32 (upper and lower), 33, 65 (upper), 80, black-and-white illustrations pp. 67, 83

Some photographs and illustrations are used by permission and are the property of the original copyright owners.

ISBN 1 56691-066-8
ISSN 1082-5134

Editors: Karen Gaynor Bleske, Don Root
Copy Editors: Nicole Revere, Deana Corbitt Shields
Production & Design: David Hurst
Cartographers: Wil Tirion, Carl Koppeschaar, Bob Race, Jason Sadler
Index: Asha Johnson

Front cover photo: Carl Koppeschaar

Distributed in the U.S.A. by Publishers Group West
Printed in Hong Kong

All rights reserved. No part of this book may be translated or reproduced in any form, except brief extracts by a reviewer for the purpose of a review, without written permission of the copyright owner.

To my son, Ralf Koppeschaar,
who does not want to settle for the Moon
but aims to be the first human to set foot on Mars.

CONTENTS

PREFACE 10

INTRODUCTION 13
  The Land 13
    The Origin of the Moon; Geological History; Geography;
    Climate; Visibility; Environmental Issues
  History and People 41
    Visionaries and Dreamers; From Firecracker to Rocket;
    Modern-Day Travelers
  The Lunar Age 63
    Clementine and the South Pole; The Discovery of Ice;
    Government

ON THE SURFACE 67
  Sightseeing Highlights 67
  Recreation 69
    Sports; Events
  Accommodations and Food 76
    Where to Stay; Where to Eat; Food and Drink
  Transportation 79
    Getting There; Getting Around
  Information and Services 85
    Officialdom; Health and Safety; Money; Communications;
    Tourist Information; What to Take; Weights and Measures

OUT AND ABOUT 94
  Moon Museum
  Attractions 97
    The Straight Wall; Alphonsus Crater; Aristarchus Crater
    and Schröter's Valley; Linné Crater; The O'Neill Bridge;
    The Mountain of Eternal Light
  The Near Side 104
    Lunox; Mare Tranquilitatis to Mare Vaporum;
    Mt. Schneckenberg; Mt. Schneckenberg to Mont Blanc Resort;
    Mont Blanc Resort; Vicinity of Mont Blanc Resort; Mont Blanc
    Resort to Schröter's Valley; Moon City; Vicinity of Moon City;
    Moon City to the Straight Wall and Mountain of Eternal Light
  The Far Side 119
    Eye on the Stars

AFTERWORD: EARTH
        WITHOUT THE MOON 126

RECOMMENDED READING 128

INDEX 131

MAPS

Ariadaeus Rille, Vicinity of 96
Bases on the Moon 68
Distance between Earth and Moon 21
Moon, The 14–15
Near Side Sights 95
Solar Eclipse 73
Terrestrial Eclipse 73
Visibility on the Moon 38

CHARTS

Apollo Moon Landings 60
How Gravity Works 93
Lunar Orbiter Program, The 60
Near Side: Capes, Mountain Ranges,
  and Mountain Peaks, The 30–31
Near Side: Oceans, Seas, Bays, Lakes, and Marshes, The 27
Near Side: Valleys, Rilles, and Scarps, The 35
Ranger Program, The 58
Russian Lunar Rovers 55
Russian Unmanned Missions to the Moon 54–55
Surveyor Program, The 58
Total Solar Eclipses on the Moon 75
Transportation on the Moon 84
Visibility on the Earth and Moon 39
Visibility on the Moon 82

SPECIAL TOPICS

Animals of the Moon 52
Death and the Moon 37
Drive Your Own Moon Rover 70
Influence of the Moon, The 19–20
Jack and Jill 40
Lunacy 112
Lunar Archives 64
Lunar Calendar 90–92
Moonbounce 120
Moon Illusion, The 123–125
Moonstruck 23
Real Moon City, The 118
Seeing Double 46
Selene 26
Father Moon 44
Soviet Mountains—A Lunar Glitch, The 56
Spin a Coin 87
Sun and Moon 99
Woman in the Moon, The 74
Words—Setting the Record Straight 61

ABBREVIATIONS

GMT — Greenwich mean time
km — kilometer
LFSO — Lunar Far Side Observatory
LM — Lunar Module
PLSS — Portable Life Support System

PREFACE

In 1856, the German sky watcher Julius Schmidt wrote:

> *In the early morning the mountains suddenly emerged from the night. Not from a grayish dawn, nor from the steaming mists of a valley, but from the deepest darkness. During this change from the long night to the day, during this dawn in a world alien to us, not a sound reaches our ears. On the Moon, the new day does not trigger the sounds of awakening animals that we are accustomed to hearing. No birds fly. No plants grace the bare soil. A deathly silence reigns on the ground and in the sky. When a boulder falls, there is not a roar. Nor do the mountains resound with an echo. The eye searches in vain for cloud formations or shadows. The sky is an inky darkness. Only the Sun, the Earth, and the stars shine.*

What a pity that Schmidt never got a chance to see real life on the Moon. How surprised he would be to hear gurgling streams and rustling bushes in the Moon's parks. Of course, humans have brought those terrestrial pleasures in spaceships. These days, in the year 2020, the Moon is bursting with life. You do have to don your moonsuit every time you go out, but once you're safely inside the domed lunar bases, you will find anything you might want. The Moon has become a second Earth, awaiting visitors. As science-fiction master Arthur C. Clarke wrote years ago in his *Interplanetary Flight* (1950):

> *There is no way back into the past; the choice, as H.G. Wells once said, is the Universe—or nothing. Though men and civilizations may yearn for rest, for the Elysian dream of the Lotus Eaters, that is a desire which merges imperceptibly into death. The challenge of the great spaces between the worlds is a stupendous one; but if we do fail to meet it, the story of our race will be drawing to its close. Humanity will have turned its backs upon the still untrodden heights and will be descending again the long slope that stretches, across a thousand million years of time, down to the shores of the primeval sea.*

With this travel guide in your pocket, you can see this New World for yourself. How about a trip to long, narrow Alpine Valley, which slices a mountain down the middle, or to the Mountain of Eternal Light? Or perhaps technology interests you, and you'd prefer a tour of the Moon's oxygen and concrete production plants. The Moon Museum documents Moon history, mythology, exploration, and settlement. Visit it first, and then hop into a Lunar Module. You'll reach your destinations in no time.

—Carl Koppeschaar, Moon City, June 2020

The trip to the Moon in Professor Calculus's spaceship in the Tintin comic book *Explorers on the Moon. The caption reads,* "Roger. We await your instructions."

INTRODUCTION

THE LAND

The barren ball we call the Moon measures 3,476 km in diameter and has a surface area of 37.5 million km^2. Compared with the Earth's surface area of roughly 500 million km^2, that may not seem like much. But don't forget that 71% of the Earth's surface is covered with water, while the Moon has no oceans, rivers, or lakes. This means that the surface area of the Moon is roughly equal to a quarter of the Earth's total land surface.

With no water, wind, or weather to erode it, the Moon's landscape is changed solely by the impacts of meteorites (although scientists have conjectured that moonquakes could also cause changes).

Thanks to those impacts, the entire Moon is covered with rock fragments varying in size from small dust particles to huge boulders dozens of meters in diameter. This so-called regolith layer averages four to five meters thick in the plains and approximately 10 meters thick in the highlands. Underneath the regolith lies a kilometer-thick layer of basalt and beneath that is a layer of stronger rock, reaching a depth of 25 km.

So many surface irregularities cover the Moon that it looks like a navigational nightmare to anyone wanting to brave its surface. Among its features are thousands of craters, kilometers of dark lava-covered plains (still incorrectly referred to as *maria,* or seas), numerous mountain ranges, isolated mountain peaks, capes (also known as promontories), valleys, scarps (cliffs), and rilles (deep clefts similar to fault lines on Earth).

THE ORIGIN OF THE MOON

The Fission Theory
Scientists have long debated how the Moon originated. In about the year 1880, Sir George Howard Darwin, the son of well-known evolutionist Charles Darwin, postulated that the Earth and the Moon once

(continues on page 18)

14 INTRODUCTION

THE LAND 15

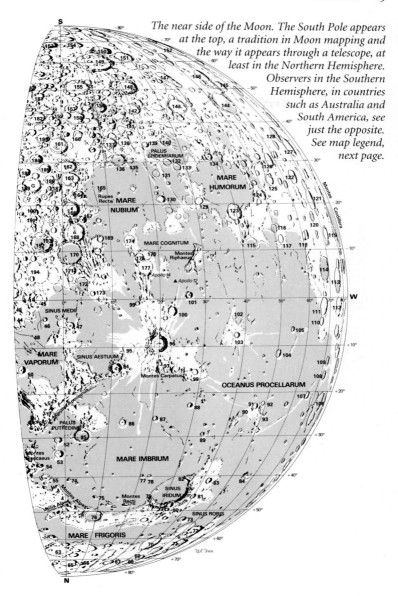

The near side of the Moon. The South Pole appears at the top, a tradition in Moon mapping and the way it appears through a telescope, at least in the Northern Hemisphere. Observers in the Southern Hemisphere, in countries such as Australia and South America, see just the opposite. See map legend, next page.

THE MOON LEGEND

1. Neper
2. Apollonius
3. Firmicius
4. Condordet
5. Taruntius
6. Picard
7. Proclus
8. Macrobius
9. Cleomedes
10. Hahn
11. Berosus
12. Gauss
13. Burckhardt
14. Geminus
15. Messala
16. Mercurius
17. Franklin
18. Cepheus
19. Oersted
20. Atlas
21. Hercules
22. Endymion
23. De La Rue
24. Vetruvius
25. Promontorium Argaeus
26. Littrow
27. Le Monnie
28. Chacornac
29. Posidonius
30. Mason
31. Plana
32. Bürg
33. Maskelyne
34. Sabine
35. Ritter
36. Arago
37. Julius Caesar
38. Plinius
39. Promontorium Archerusia
40. Menelau
41. Bessel
42. Linné
43. Godin
44. Agrippa
45. Rhaeticus
46. Triesnecker
47. Pallas
48. Hyginus
49. Boscovich
50. Manilius
51. Conon
52. Autolycus
53. Aristillus
54. Theaetetus
55. Cassini
56. Callipus
57. Alexander
58. Eudoxus
59. Aristoteles
60. Gärtner
61. Arnold
62. Meton
63. W. Bond
64. Barrow
65. Goldschmidt
66. Anaxagoras
67. Philolaus
68. Anaximenes
69. Carpenter
70. J. Herschel
71. Pythagoras
72. Babbage
73. Harpalus
74. Mons Piton
75. Mons Pico
76. Plato
77. Le Verrier
78. Helicon
79. Promontorium Laplace
80. Bianchini
81. Sharp
82. Promontorium Heraclides
83. Mairan
84. Mons Rümker
85. Archimedes
86. Timocharis
87. Lambert
88. Euler
89. Delisle
90. Prinz
91. Aristarchus
92. Herodotus
93. Vallis Schröteri
94. Eratosthenes
95. Stadius
96. Copernicus
97. Gay Lussac
98. Mayer
99. Gambart
100. Reinhold
101. Landsberg
102. Encke
103. Kepler
104. Marius
105. Reiner
106. Otto Struve
107. Seleucus
108. Krafft
109. Cardanus
110. Cavalerius
111. Hevelius
112. Hedin
113. Riccioli
114. Grimaldi
115. Letronne
116. Billy
117. Hansteen
118. Sirsalis
119. Rocca
120. Crüger
121. Darwin

122 Byrgius	165 Birt	208 Azophi
123 Gassendi	166 Arzachel	209 Abenezra
124 Mersenius	167 Alpetragius	210 Geber
125 Cavendish	168 Alphonsus	211 Tacitus
126 Vieta	169 Davy	212 Almanon
127 Lagrange	170 Ptolomaeus	213 Abulfeda
128 Piazzi	171 W. Herschel	214 Fabricius
129 Agatharchides	172 Flammarion	215 Janssen
130 Bullialdus	173 Mösting	216 Metius
131 Kies	174 Guericke	217 Brenner
132 Mercator	175 Parry	218 Rheita
133 Campanus	176 Bonpland	219 Neander
134 Vitello	177 Fra Mauro	220 Piccolomini
135 Hesiodus	178 Manzinus	221 Fracastorius
136 Pitatus	179 Jacobi	222 Beaumont
137 Gauricus	180 Cuvier	223 Catharina
138 Würzelbauer	181 Licetus	224 Cyrillus
139 Cichus	182 Stöfler	225 Theophilus
140 Capuanus	183 Nonius	226 Mädler
141 Heinsius	184 Aliacensis	227 Ididorus
142 Wilhelm	185 Werner	228 Capella
143 Mee	186 Blanchinus	229 Torricelli
144 Schickard	187 La Caille	230 Hypatia
145 Wargentin	188 Apianus	231 Zöllner
146 Phocylides	189 Playfair &	232 Delambre
147 Schiller	Playfair G	233 Furnerius
148 Longomontanus	190 Airy	234 Stevinus
149 Clavius	191 Argelander	235 Snellius
150 Blancanus	192 Albategnius	236 Reichenbach
151 Scheiner	193 Klein	237 W. Humboldt
152 Bailly	194 Hipparchus	238 Petavius
153 Curtius	195 Vlacq	239 Santbech
154 Moretus	196 Hommel	240 Colombo
155 Maginus	197 Pitiscus	241 Goclenius
156 Tycho	198 Baco	242 Gutenberg
157 Saussure	199 Barocius	243 Messier &
158 Orontius	200 Maurolycus	Messier A
159 Nasireddin	201 Buch	244 Vendelinus
160 Lexell	202 Büsching	245 Langrenus
161 Walter	203 Riccius	246 Ansgarius
162 Regiomontanus	204 Rabbi Levi	247 La Pérouse
163 Purbach	205 Zagut	248 Kästner
164 Thebit	206 Pontanus	249 Gilbert
	207 Sacrobosco	

Astrologers note the phase of the Moon while a baby is being born. No connection has been found between the Moon and human behavior. Though it may be pure coincidence, the female menstrual cycle does average the same number of days as the lunar month.

were one. According to his theory, when the Earth was still a red-hot, semi-liquid ball of lava, it spun very quickly on its axis—approximately once every five hours. This rapid rotation made it unstable, so that a bulge formed around the equator. At some point this bulge broke off from the Earth, forming the Moon. Later astronomers postulated that if Darwin's theory were true, this could not have happened without leaving an enormous scar in the Earth's crust. And the Earth has just the kind of depression that would fit the bill: the giant basin of the Pacific Ocean.

THE INFLUENCE OF THE MOON

IN NEW GUINEA AND SOME PARTS OF AFRICA WOMEN HOLD UP THEIR BABIES TO the New Moon so that they will grow up to be healthy and strong. In Greenland women believe that they are more likely to conceive if they sleep in the moonlight. Indian tribes along the Orinoco bury a smoldering fire in the ground during a lunar eclipse, since they think that when the Moon "goes out," all the fire on Earth will also be extinguished. To save the fire, it has to be hidden from the face of the Moon. A Crescent Moon on the increase is good luck for travelers and lovers, and a crescent-shaped pendant is considered an amulet for someone on the road.

Primitive superstition? How many people believe crops should be planted by the light of a New Moon? And many a lumberjack swears that the quality of the wood is better if a tree is felled during a waning Moon. It's easy to dismiss such beliefs as superstitious nonsense, but first let's take a closer look at the Moon's influence on the Earth.

Geocosmic Influences

The Sun clearly influences the weather. After all, solar heat is the engine that regulates our terrestrial atmosphere and seawater. But what about the Moon, which is 400 times closer to the Earth? The answer is that the Moon radiates so little heat that it doesn't affect the Earth in the slightest.

Even the tidal action between the Earth and the Moon is far too weak to influence the weather in any significant way. The difference in pressure between the two is a negligible 0.03 to 0.07 millibar. To give you an idea of what this means, the normal pressure difference between high and low pressure areas is 40 millibars, or roughly a thousand times higher. Consequently, for a long time scientists doubted whether the Moon had any influence on the weather at all.

However, in 1960 scientists found a connection between the phases of the Moon and the rainfall in New Zealand: Around New Moon, less rain fell than at Full Moon. A similar connection was found in other countries, though it varied from place to place. For example, American scientists worked their way through enormous amounts of precipitation data and discovered that the amount of rain was at its maximum a few days after Full Moon and New Moon. The amount of sunshine during these periods was at a minimum. These connections are statistically significant, which is another way of saying that they can't be attributed to coincidence.

continues

Still, it is necessary to treat such results with caution. We know, for example, that on the Dutch island of Schiermonnikoog, the rainfall does seem to be influenced by the Moon. However, this influence can be attributed to the tides. The mudflats are dry during an ebb tide. Because the Sun warms the sand faster than it does the seawater, a warm layer of air is formed faster above the dry land than above the sea. This warm air rises and helps to form clouds. And so it happens that the changes of ebb and flow, and therefore also indirectly the phases of the Moon, are reflected in the precipitation data.

Marine Animals

The cyclic behavior of living creatures is influenced to a certain degree by moonlight. For example, the best time for oysters to spawn in Holland's East Scheldt River is between 26 June and 10 July in the 24-hour period after Full or New Moon. In Europe, adult eels wait for a waning Moon before making their long and necessary journey to the Sargasso Sea to spawn. The *palolo* worm, which burrows in the coral reefs of tropical waters, always reproduces in October or November when the Moon goes into its last quarter. In Samoa and the Fiji Islands, where the animal is considered a delicacy, the arrival of the *palolo* worm is a high point in the local calendar.

Humans

If moonlight influences the behavior of marine animals, does it also influence that of human beings? Humans appear to have a day-night rhythm of a little more than 24 hours, which is adjusted daily under the influence of sunlight. Other rhythms have never been found (see the special topic "Lunacy" in the Out and About chapter), although according to Dr. Piet Jongbloet of the Netherlands, the female menstrual cycle, occurring as it does an average of 29.5 days, coincides remarkably with the average lunar cycle (the time it takes the Moon to revolve around the Earth). And the average pregnancy lasts 266 days, or nine lunar cycles. There may be a connection. Or it may be a relic of a prehistoric period, when a person's chance of survival (especially that of a vulnerable newborn) was directly related to a variety of natural phenomena such as vegetation, temperature, and the amount of daylight or moonlight. This has absolutely nothing to do with astrology.

Darwin's theory can't be true, based on evidence gathered from Moon rocks. The lunar surface is as old as the Earth's, and the ocean floor is much younger. Yet for a long time, the evidence appeared to support the fission theory. Astronomers generally agree that the Moon was once much closer to Earth than it is now. But it's the action of tides, not fission, that explains why the Moon is moving away from the Earth.

Longer Days
The gravitational attraction between the Earth and the Moon—which on Earth manifests itself in the tides—acts as a brake on the rotation of both Earth and Moon. The Earth's spin has gradually slowed; its 24-hour day is getting still longer. The smaller Moon has been slowed down even more. It now rotates once every 27 days, which is how long it takes the Moon to complete one revolution of the Earth. This is why the same side of the Moon is always facing the Earth. It's as if the Earth has locked the Moon into position.

As the rotation of the Earth slows, the Moon is pushed farther away so that the total angular momentum of the system remains constant. (Think of figure skaters who spin slowly with their arms extended and speed up when they pull them toward their bodies.) The Moon is moving away at the rate of 3.5 cm a year and now is about 384,400 km distant. At 3.5 cm per year, in 100 million years it will have moved 3,500 km farther away from Earth.

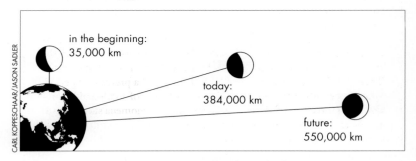

Distance between Earth and Moon: Because of the tidal friction between the Earth and the Moon, the Earth is spinning more slowly around its axis. The Moon is gradually moving away from the Earth.

As the Moon-Earth distance increases, the days on Earth will get longer. This will continue until four billion years from now, when the Earth too will be locked into position. At the end, the Earth will continue to show the same side to the Moon. One day on Earth will then be as long as the month (one revolution of the Moon around the Earth). The Moon will hover motionlessly just above the Earth's equator. It will no longer come up or go down. Those people living on the side of the Earth turned away from the Moon will never see the Moon again.

The Giant-Impact Theory
Current data indicate that the Moon was created after a celestial object the size of the planet Mars collided with the Earth. The crash must have occurred in the solar system's early days, when large chunks of debris were still floating around.

According to the giant-impact (or big-whack) theory, when the "invader" grazed against the primitive Earth, its mantle and part of the Earth's upper mantle pulverized. The collision hurled this matter into an orbit around the Earth, briefly surrounding the planet with a ring. But the individual particles in the ring soon began attracting each other, massing into a celestial object that grew and grew until it finally swept up all the ring's smaller particles. The Moon has been revolving around the Earth ever since.

GEOLOGICAL HISTORY

In all likelihood, during the first 200 million years of its existence, the Moon was entirely molten. However, the outer layer cooled and solidified approximately 4.4 billion years ago. Over the next couple of hundred million years, it was heavily bombarded with fragments from the solar system.

Owing to the radioactive decay of the uranium, potassium, and thorium in the Moon's core, the mantle directly under the crust liquefied. This magma bubbled to the surface 3.1-3.9 billion years ago, filling the largest crater basins with lava and creating the *maria,* which are often circular. After that, the Moon experienced almost no dramatic changes, except the arrival of a few large meteorites. The most striking of them created impact craters such as Copernicus, Aristarchus, and Tycho.

MOONSTRUCK

For a moment there in 1991, the Moon was in danger of being smashed to bits, and not by a giant meteorite: Iowa State University mathematics professor Alexander Abian proposed that we blow up the Moon.

Abian's intent was to eliminate bitter winters and scorching summers. He thought that if the Moon were blasted to smithereens, some of its pieces would land on Earth and straighten the 23° tilt of the Earth's axis. The amount of sunlight striking every part of the untilted globe would not vary from summer to winter.

"Perpetual spring!" Abian exclaimed. Earth would have a pleasant year-round climate without cyclones, typhoons, hurricanes, tornadoes, and other storms caused mainly by sharp temperature differences.

Fortunately, Abian met with some difficulties in persuading the skeptical world to sacrifice the Moon and tamper with the Earth's climate. Biologists in particular opposed his ideas. Indeed, penguins and polar bears would not survive in Abian's iceless world. Neither would the *palolo* worms of the Pacific Islands. They rely on the Moon for cues about when to spawn.

And what about humankind? Some scientists argue that even our fertility cycle is still somewhat influenced by the revolutions of the Moon. If we had carried out Abian's proposal, we might have experienced an ominous modern application of an old English proverb about the short life span of babies born on moonless nights: "No Moon, no man."

Giordano Bruno Crater

On the evening of 25 June 1178, five English monks witnessed what was presumably a meteorite striking the Moon. "Now there was a bright new moon, and as usual in that phase its horns were tilted toward the east," reads the report in the *Chronicle of Gervase of Canterbury*. "And suddenly the upper horn split in two. From the midpoint of this division a flaming torch sprang up, spewing out, over a considerable distance, fire, hot coals, and sparks."

The image of a torch spewing fire, hot coals, and sparks describes the impact of a meteorite fairly accurately. Such a recent impact should have left traces in the form of a fresh-looking ray crater. And lo and behold, a newish crater known as Giordano Bruno lies at the edge of the far side of the Moon, just where the monks said it was.

The American Apollo program provided a second indication that a large impact occurred on the Moon eight centuries ago. When a small object traveling at high speed strikes a larger object, the force of the impact causes the larger object to oscillate for a while. Though the oscillation gradually fades, NASA scientists reasoned that the Moon should still reveal traces of an oscillation started by an impact 800 years ago. To test their hypothesis, the scientists designed the lunar laser-ranging technique: Apollo astronauts placed mirrors on the Moon that were used to reflect laser beams emitted from Earth. By measuring the time it took a laser pulse to travel from the Earth to the Moon and back again, scientists were able to determine the distance between the two with a high degree of accuracy.

Those measurements revealed that the Moon oscillates with a period of three years and with a deviation of three meters—results absolutely consistent with the hypothesis that the Giordano Bruno Crater was formed about eight centuries ago by the impact of a meteorite.

GEOGRAPHY

Astronomers have developed a nomenclature for the foremost morphological features of the Moon. The names of most of the craters are based on the system developed by Giambattista Riccioli, an Italian Jesuit who produced a fairly good map of the Moon in 1651. Riccioli named the most prominent craters after famous scholars, such as Tycho, Copernicus, and Ptolemy. This system has been maintained and expanded, so that at present the International Astronomical Union has accepted 1,843 names. (Notwithstanding Riccioli's system, some craters have been named after astronauts.) Since the mountain ranges are normally named after terrestrial mountain ranges, don't be surprised to find the Moon dotted with the Apennines, the Alps, the Caucasus, the Carpathians, and the Pyrenees. The "oceans," "seas," "bays," "lakes," and "marshes" have somewhat more romantic names. You might run across Oceanus Pro-

Giordano Bruno, a Dominican priest, was burned at the stake in Rome in 1600 for heresy. A large crater, probably resulting from the impact of a meteorite on the far side of the Moon in 1178, has been named after him.

> ## SELENE
>
> IN ANCIENT GREECE, SELENE, THE SISTER OF THE SUN GOD HELIOS, WAS THE Moon goddess. Every evening she drove across the sky in a chariot drawn by two white horses. Selene fell in love with the beautiful shepherd Endymion, the son of a king from Elis, and bore him 50 children. Every night she kissed him while he lay sleeping. Zeus endowed the prince with eternal youth, at the same time plunging him into an eternal sleep on Mt. Latmos. Still, the faithful Selene comes to visit him every night in her chariot, pausing to kiss the sleeping Endymion, the setting Sun. Her name has been given to the branch of astronomy dealing with the Moon: selenology.
>
> In ancient Greece, all cloven-hoofed animals were dedicated to Selene, because their hooves were shaped like "Cs," the symbol for Selene in early Greek script. That's why hooved animals were sacrificed at New Moon festivals.
>
> In medieval art the Virgin Mary was often represented enthroned on the Moon, just as Selene had been before her. She is often referred to as The Moon of the Church and Our Moon and is said to control the Moon, stars, planets, seas, and tides. Her name derives from the Latin word for sea, *mare*, and she always wears blue, which represents the sea and the sky.

cellarum ("Ocean of Storms"), Mare Crisium ("Sea of Crises"), Sinus Aestuum ("Seething Bay"), Lacus Somniorum ("Lake of Dreams"), and Palus Nebularum ("Marsh of Mists"). You will find the Latin names and the English equivalents of all these surface features in the charts.

Seas

Even without a telescope, you can see from Earth that the Moon is covered with light and dark spots. The dark patches are the areas that Galileo Galilei called *maria* (singular *mare*, Latin for sea). The imagination of the ancient selenographers was so sparked by the thought of vast expanses of water on the Moon that even after it was proved that the Moon had no water, they continued to refer to the dark spots as seas. The tradition persists.

Nowadays we know that these *maria* are in fact lava deserts, drier and more desolate than any desert on Earth. Unlike terrestrial deserts, they have no oases, unless you count the human-made lunar bases. Nor will

THE NEAR SIDE: OCEANS, SEAS, BAYS, LAKES, AND MARSHES

OCEANUS PROCELLARUM	OCEAN OF STORMS
MARE ANGUIS	SERPENT SEA
MARE AUSTRALE	SOUTHERN SEA
MARE COGNITUM	KNOWN SEA
MARE CRISIUM	SEA OF CRISES
MARE FECUNDITATIS	SEA OF FERTILITY
MARE FRIGORIS	SEA OF COLD
MARE HUMBOLDTIANUM	HUMBOLDT'S SEA
MARE HUMORUM	SEA OF MOISTURE
MARE IMBRIUM	SEA OF SHOWERS
MARE MARGINIS	BORDER SEA
MARE NECTARIS	SEA OF NECTAR
MARE NOVUM*	NEW SEA
MARE NUBIUM	SEA OF CLOUDS
MARE ORIENTALE	EASTERN SEA
MARE PARVUM*	LITTLE SEA
MARE SERENITATIS	SEA OF SERENITY
MARE SMYTHII	SMYTH'S SEA
MARE SPUMANS	FOAMING SEA
MARE STRUVE*	STRUVE'S SEA
MARE TRANQUILITATIS	SEA OF TRANQUILITY
MARE UNDARUM	SEA OF WAVES
MARE VAPORUM	SEA OF VAPORS
SINUS AESTUUM	SEETHING BAY
SINUS IRIDUM	BAY OF RAINBOWS
SINUS LUNICUS	BAY OF LUNA
SINUS MEDII	CENTRAL BAY
SINUS RORIS	BAY OF DEWS
LACUS AESTATIS	SUMMER LAKE
LACUS AUTUMNI	AUTUMN LAKE
LACUS MORTIS	LAKE OF DEATH
LACUS SOMNIORUM	LAKE OF DREAMS
LACUS VERIS	SPRING LAKE
PALUS EPIMEDIARUM	MARSH OF EPIDEMICS
PALUS NEBULARUM	MARSH OF MISTS
PALUS PUTREDINIS	MARSH OF DECAY
PALUS SOMNI	MARSH OF SLEEP

* ON HISTORICAL MAPS

Mare Imbrium seen from above. The dark and nearly flat crater underneath it is Plato. To the left of Plato, near the dividing line between day and night, is the famous Alpine Valley.

you find windswept sand dunes—since the Moon has no atmosphere, there isn't so much as a breeze anywhere on the Moon.

The *maria* take up a large part of the half of the Moon visible from Earth. Oceanus Procellarum is the largest of these lunar seas. It's approximately 3,000 km wide and irregular in shape. Other lunar seas, such as Mare Imbrium, Mare Serenitatis, and Mare Crisium, are more or less round, presumably formed by the impacts of enormous asteroids or comets. The impacts were apparently so great that they caused molten lava to ooze out of the Moon and flow over the surrounding surfaces. "Ghost craters"—depressions still recognizable under the lava layer—were created where the lava flowed over craters that had already existed on the Moon's surface. A few of the craters are only partially covered.

The *maria* are deeper in the middle than at their edges with respect to the average reference diameter of the Moon. Mare Imbrium is seven km deep in the middle. Mare Crisium has a depth of four km, while Oceanus Procellarum is a relatively shallow one km deep at its center. Highlands and mountain ranges surround most of the *maria*. With the exception of the relatively small Mare Moscoviense ("Moscow Sea"), which was photographed by the Russian Luna 3, the far side of the Moon has no seas.

Mountains

The mountains on the Moon top out at about the same heights as their counterparts on Earth. Epsilon Peak in the South Pole, for example, measures 9,050 meters from top to bottom, which is a bit taller than Mt. Everest. In the Rook Mountains, some of the peaks are nearly 7,000 meters, while Mt. Huygens in the Apennines rises to a majestic 5,400 meters.

Unlike the Earth, which has real, liquid seas and therefore a sea level, the Moon doesn't have a uniform base level from which to measure. Heights are instead measured relative to the surrounding terrain. Given that the Moon's diameter is a quarter of the Earth's, you might claim that these peaks are, relatively speaking, four times higher than on Earth.

But aside from their height, the Moon's mountains do not greatly resemble those on Earth. They have not been formed by the same processes as on Earth, so you won't find any long, steeply upthrust ranges. Instead, mountains here consist of a series of isolated peaks unrelated to each other. Besides that, they have little topographical relief; in fact, none of the mountain peaks on the Moon is pointed.

The mountainous terrain in the south. You can clearly see the Clavius walled plain (top center), the Tycho ray crater (directly below it) and the Straight Wall (bottom center).

The classical method of determining the height of mountains on the Moon was to measure the length of the shadows they cast. But these shadows were sometimes so deceptively pointed that some of the ancient selenographers began to describe them in terms of "peaks" and "needles."

Modern measurement techniques provide a very different picture. For example, though Epsilon Peak is 9,050 meters high, it measures 50 km at its base, which comes down to a height-baseline ratio of 1:5.6. The average slope therefore amounts to 21°. For the other peaks in the same mountain range, this ratio turns out to be 1:7 and 1:8, or a slope of 15°, which can easily accommodate a vehicle.

Even the peaks whose shadows appear to be quite pointed, such as the isolated Pico and Piton Capes in the western part of Mare Imbrium, are actually relatively flat. For example, the southernmost of the

THE NEAR SIDE: CAPES, MOUNTAIN RANGES, AND MOUNTAIN PEAKS

CAPE (PROMONTORIUM)

	LAT.	LONG.	LOCATION
AENARIUM	19°S	8°W	MARE NUBIUM, NEAR STRAIGHT WALL
AGARUM	15°N	65°E	MARE CRISIUM, NEAR CONDORCET
AGASSIZ	42°N	2°E	MARE IMBRIUM, NEAR CASSINI
ARCHERUSIA	17°N	22°E	MARE SERENITATIS, EASTERN SIDE OF HAEMUS MTS.
BANAT	17°N	26°W	CARPATHIAN MTS., NEAR TOBIAS MAYER
DEVILLE	46°N	0°	MARE IMBRIUM, NEAR CAPE AGASSIZ
FRESNEL	29°N	5°E	MARE SERENITATIS, END OF APENNINES
HERACLIDES	40°N	34°W	WESTERN CAPE OF SINUS IRIDUM
KELVIN	27°S	33°W	MARE HUMORUM, NEAR HIPPALUS
LAPLACE	46°N	26°W	SINUS IRIDUM
LAVINIUM	14°N	48°E	MARE CRISIUM, PART OF "O'NEILL BRIDGE"
OLIVIUM	15°N	48°E	MARE CRISIUM, PART OF "O'NEILL BRIDGE"
PICO	46°N	9°W	MARE IMBRIUM, OPPOSITE ALPINE VALLEY
PITON	41°N	1°W	MARE IMBRIUM, OPPOSITE ALPINE VALLEY
RÜMKER	41°N	58°W	SOUTH OF SINUS RORIS

MOUNTAIN RANGES (MONTES):

ALPES	ALPS, NORTHERN BORDER OF MARE IMBRIUM
APENNINUS	APENNINES, SOUTHWESTERN BORDER OF MARE IMBRIUM
CARPATUS	CARPATHIAN MTS., SOUTHERN BORDER OF MARE IMBRIUM
CAUCASUS	CAUCASUS MTS., BETWEEN MARE SERENITATIS AND PALUS NEBULARUM
CORDILLERA	CORDILLERA MTS., WESTERN EDGE OF MOON, NEAR GRIMALDI
D'ALEMBERT*	D'ALEMBERT MTS., WESTERN EDGE OF MOON
DOERFEL*	DOERFEL MTS., SOUTH-SOUTHWESTERN EDGE OF MOON
HAEMUS	HAEMUS MTS., SOUTHERN BORDER OF MARE SERENITATIS
HARBINGER	HARBINGER MTS., BETWEEN MARE IMBRIUM AND OCEANUS PROCELLARUM
JURA	JURA MTS., NORTHEASTERN BORDER OF MARE IMBRIUM
LEIBNITZ*	LEIBNITZ MTS., SOUTHERN EDGE OF MOON
PYRENNAEUS	PYRENEES, EAST OF MARE NECTARIS
RECTI	MARE IMBRIUM, SOUTHWEST OF PLATO
RIPHAEUS	RIPHEAN MTS., BETWEEN MARE COGNITUM AND OCEANUS PROCELLARUM
ROOK	ROOK MTS., AROUND MARE ORIENTALE
SPITZBERGENSIS	SPITZBERGEN MTS., MARE IMBRIUM, NORTH OF ARCHIMEDES
TAURUS	TAURUS MTS., EAST OF MARE SERENITATIS
TENERIFFE	TENERIFFE MTS., MARE IMBRIUM, SOUTH OF PLATO
URAL*	URAL MTS., EXTENSION OF RIPHEAN MTS.

MOUNTAIN PEAKS (MONS)

	LAT.	LONG.	HEIGHT	LOCATION
AMPÈRE	20°N	4°W	3,200 M	APENNINES
BRADLEY	22°N	1°E	3,500 M	APENNINES
HADLEY	27°N	5°E	3,500 M	APENNINES
HUYGENS	20°N	3°W	5,400 M	APENNINES
LA HIRE	28°N	25°W	1,500 M	MARE IMBRIUM
SERAO	17°N	6°W	2,000 M	APENNINES
WOLFF	17°N	7°W	3,500 M	APENNINES
MONT BLANC	45°N	1°E	3,600 M	ALPS

* ON HISTORICAL MAPS

three mountains forming Piton, whose combined bases amount to 20 km, rises no higher than 2,250 meters above the surrounding sea.

Craters
The most striking features on the Moon are its craters. The largest of these, approximately 100-300 km in diameter, are known as **walled plains.** Those having diameters of 20-100 km are known as **ring plains.** Formations smaller than that are simply labeled **craters.** And craters with diameters smaller than one kilometer are known as **crater pits** or **craterlets.**

Bailly, the largest walled plain, is 295 km in diameter and four km deep. Although Bailly is so far from Moon settlements that visitors rarely see its impressive dimensions, travelers to the Mountain of Eternal Light could make a short side trip to the crater, which lies near the edge of the far side of the Moon.

Clavius, with a diameter of 232 km and a depth of five km, lies between Moon City and the South Pole. It's smaller than Bailly, but on Earth it would take up a region about the size of Massachusetts or Costa Rica.

Most of the round formations on the Moon are bound to be ring plains and craters. They have been better preserved than the larger walled plains. Sometimes they are lined with "terraces," concentric rings of collapsed walls separated by ravines. Unlike walled plains, ring plains and craters often have a central mountain peak rising from the crater floor. These central peaks are not really mountains, but freakish rock formations. Some of them tower as much as 1,800 meters above the crater floor.

Copernicus, with a diameter of 90 km, is the most beautiful ring plain on the Moon. Its walls form an unbroken ring that rises four km above the crater floor. The ringwall descends on its outer face in gently sloping rings. At the foot are many scarps and clefts that radiate nearly 100 km from the crater. The inner ringwall is steeper and descends to the floor in steps.

Copernicus is also a so-called **ray crater.** At Full or a nearly Full Moon, even Earth-bound observers can see how the rays of ejected material radiate hundreds of kilometers from the crater. However, **Tycho** boasts the most beautiful rays. Like Copernicus, Tycho is hard to miss during Full Moon. Tycho's rays spread out more than 1,500 km, stretching all the way from its outer walls to the edge of the near side of the Moon.

(above) The Lunar Transfer Vehicle departs from a space station in a low Earth orbit. Most visitors stay overnight in the Hilton Space inside the station before continuing their journeys to the Moon. (below) The Lunar Transfer Vehicle arrives at the space station orbiting the Moon. (Illustrations: Pat Rawlings/NASA)

THE LAND 33

One of the most beautiful walled plains is Langrenus, named after the Dutch mathematician and astronomer Michiel Florentius van Langren. Langrenus is 136 km in diameter. The terraced walls rise 2.7 km above the lowest point in the plain.

All kinds of theories have been put forward to explain the existence of craters on the Moon. Initially, scientists hypothesized that huge volcanic eruptions had created them. Although the craters were much larger than could normally be accounted for by volcanic activity, the scientists attributed this to the Moon's lower gravity.

These days, astronomers agree that all craters larger than 20 km were created when rock blocks from outer space crashed into the lunar soil. The explosive energy that would be released upon such an impact is the only thing large enough to account for craters that size. For example, if a rock block with a radius of one kilometer were to hit a moon or a

(opposite) Once the landing craft has descended to 500 meters, it is guided to the landing platform near the lunar base by laser. (Pat Rawlings/NASA)

planet at a velocity of 20 km per second, the force of the resulting explosion would equal half a billion megatons of TNT. Like that of an atom bomb, the explosion forms a saucerlike crater with a blanket of ejected material around it. The rock fragments scattered by the force of the explosion form secondary impact craters.

The larger craters collapse, forming stepped ringwalls. Occasionally a mountain peak rises in the middle because the compressed crust attempts to regain its equilibrium (isostasy). Inside some of the larger craters, magma flows create a relatively level lava bed.

The early scientists weren't all wrong, though; a certain kind of vulcanism does occur on the Moon. The *maria* and some of the walled and ring plains are strewn with **dome-top craters** and **crater cones** that tower above the crater floor. The dome-top craters, which look a lot like terrestrial shield volcanoes or the volcanic lakes in Germany's Eifel Mountains, are indeed volcanic and occur predominantly at the edges of lunar seas. Crater cones are cone-shaped hills with small openings. They are similar to the tiny tuff rings in terrestrial volcanoes and are caused by gas emissions in the lunar soil.

Valleys, Rilles, and Scarps

Valleys and rilles are fault regions shaped like riverbeds. The term valley is used to refer to wider regions, and the word rille to narrower ones.

The best-known valleys are Alpine Valley southeast of Plato and Schröter's Valley northwest of Aristarchus. Two permanently inhabited lunar bases have been built near these valleys.

Rilles, or deep clefts, fall into two types: ordinary rilles and crater rilles. The ordinary rilles are the most plentiful. The most eye-catching is Ariadaeus Rille, a 300-km-long, three- to five-km-wide, and 800-meter-deep chasm connecting Mare Tranquilitatis with Mare Vaporum.

Crater rilles are narrow channels linking rows of tiny, unwalled, nearly contiguous craters known as chain craters. One of the most beautiful crater rilles lies near Hyginus Crater.

Scarps are faults in the Moon's crust. The most stunning example is the Straight Wall ("Rupes Recta") between Moon City and the South Pole.

Rocks

The lava in the lunar *maria* has oozed out of fissures and faults in the

THE NEAR SIDE: VALLEYS, RILLES, AND SCARPS

VALLEYS (*VALLIS*):

ALPES	ALPINE VALLEY, SOUTHEAST OF PLATO
BAADE	EAST AND SOUTHEAST OF BAADE
INGHIRAMI	NORTHWEST OF INGHIRAMI
PALITZSCH	ALONG EASTERN WALL OF PETAVIUS
RHEITA	EAST OF METIUS, FABRICIUS AND JANSSEN
SCHRÖTERI	SCHRÖTER'S VALLEY, NORTHWEST OF ARISTARCHUS
SNELLIUS	BETWEEN SNELLIUS AND STEVINUS
TAURUS-LITTROW	TAURUS-LITTROW REGION

RILLES (*RIMA*):
(PARTIAL LISTING)

ARIADAEUS	BETWEEN MARE TRANQUILITATIS AND MARE VAPORUM
CAUCHY	NEAR CAUCHY IN MARE TRANQUILITATIS
HYGINUS	MARE VAPORUM
STADIUS	BETWEEN COPERNICUS AND ERATOSTHENES

SCARPS (*RUPES*):

ALTAI	SOUTHWESTERN BORDER OF MARE NECTARIS, NEAR PICCOLOMINI
CAUCHY	SOUTH-SOUTHWEST OF CAUCHY
KELVIN	NEAR CAPE KELVIN IN MARE HUMORUM
LIEBIG	NEAR LIEBIG IN MARE HUMORUM
MERCATOR	NEAR MERCATOR BETWEEN MARE NUBIUM AND PALUS EPIDEMARIUM
RECTA	STRAIGHT WALL IN MARE NUBIUM

ground. It must have been very thin to have flowed over such great distances. The minerals found in the lava are the same as those found in terrestrial basalt: pyroxene, plagioclase, olivine, spinel, and ilmenite. Seven new minerals have also been found, including pyroxferroite and Armalcolite (named for the Apollo 11 crew: ARMstrong, ALdrin, and COLlins). These lunar basalts contain much more titanium and iron than terrestrial basalts and are rich in oxygen compounds. They also contain 10 times less sodium and potassium than rocks on Earth and do not contain any water.

Scientists examine a Moon rock.

Some of the mountainous highlands on the Moon consist of more than 80% plagioclase. This igneous rock has twice as much calcium and aluminum as the *mare* basalts, but less titanium, magnesium, and iron. Rocks rich in potassium, uranium, and thorium have also been found.

CLIMATE

Temperature
Because the Moon has no atmosphere, it experiences enormous temperature swings. At the equator, the temperature at night drops to -173° C and rises during the day to +117° C. During lunar eclipses, in which the Earth may block the sunlight for up to two and a half hours, the temperature drops all of a sudden to about -100° C.

Travel Seasons
The Moon has no true seasons because its axis is inclined so slightly that the Sun is at the same height in the sky at every latitude. At or near the equator, the Sun is almost invariably at its zenith. Conversely, around the poles the Sun is always on the horizon during the day. Each lunar zone therefore experiences only one "season"—spring, summer, fall, or win-

DEATH AND THE MOON

ANCIENTS BELIEVED THAT THE MOON WAS EITHER A HOME OR A STOPPING place for the dead. This concept may come from the cyclical nature of the Moon or its eerie appearance. In the Indian Upanishads, it's written that the Moon is just a detour on the way to death. But spirits eventually return to Earth in the form of falling rain or semen. The following legend is told at the Pitcher Fourth festival:

> *There once was a sister who had many brothers. On Pitcher Fourth day, a time when women fast to attain longevity for their husbands, the sister was fasting. Her youngest brother felt so sorry for her that he scaled a tree and fooled his sister into thinking the Moon had risen by placing a light in the tree. When she saw it, she ended her fast, and her husband died. She guarded his body until the next Pitcher Fourth day. She then cut herself and emptied the blood into her husband's mouth. He regained life at once, his spirit having lain dormant on the Moon.*

In North Australia, the people of Melville Island link adultery and death with the Moon in this tale:

> *Purakapali was hunting while his family remained in their village. While he was gone his wife, Pima, went off with her lover Tjapara, the Moon, leaving her son alone.*
>
> *When Purakapali returned, he found his son dead. He was enraged at his wife and Tjapara. To placate the grieving father, Tjapara said, "Give me your dead son, and in three days I will revive him." But Purakapali was too upset to trust him. He quarreled and fought with Tjapara and eventually killed him.*
>
> *Purakapali took his son and walked into the sea, shouting, "As my son has died and will never return, so shall all human life." And, true to his claims, Purakapali never returned. But Tjapara, the Moon, was back in three days.*

ter. Since the Portable Life Support System (PLSS) completely regulates the temperature inside your space suit, and the domes of Moon City and other settlements are kept at a comfortable room temperature, the wardrobe you bring, from Bermuda shorts to formal dress, depends entirely on your frame of mind. (See "What to Take" in the On the Surface chapter.)

VISIBILITY

Observers standing on the lunar surface usually find their view hampered by the short horizon. For instance, seen from above, the 232-km crater Clavius is awesome. But standing in the middle of it would provide a less-impressive vista—the ringwall would be out of sight below the horizon. Since the Moon is much smaller than the Earth, its surface curvature is more pronounced. When you're standing on flat ground, the horizon is a mere 2.5 km away. And the Moon lacks an atmosphere. On Earth, the atmosphere deflects light rays, and under ideal circumstances you can see far beyond the theoretical horizon. At high altitudes this refraction decreases because the air thins out, so that the theoretical and actual visible horizons on the Earth eventually approach each other.

ENVIRONMENTAL ISSUES

Astrotrash

In the early years of the space race, the Russians and Americans both crash-landed numerous spacecraft on the Moon as they strove to be the first to land a man there. The remains of these early Lunokhods, Rangers, and Lunar Orbiters lie scattered about the Moon, though most of them have been picked over by scientists and visitors eager for souvenirs. (See "Modern-Day Travelers" in the "History and People" section.)

And when humans finally did set foot on the Moon, they neglected to clean up after themselves when they left. After completing the first manned exploration of the lunar surface, Neil Armstrong and Buzz Aldrin used the Eagle's landing gear as a lift-off platform to get back to

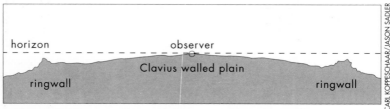

Visibility on the Moon: To the observer standing in the middle of the Clavius walled plain, the ringwall surrounding the plain lies below the horizon.

VISIBILITY ON THE EARTH AND MOON

	Moon	Earth		
Height	Visible Horizon (km)	Theoretical Horizon (km)	Refraction (arcminutes)	Visible Horizon (km)
150 cm	2.3	4.4	34	67.5
175 cm	2.5	4.7	34	67.8
2 m	2.6	5.1	34	68.1
3 m	3.2	6.2	34	69.3
4 m	3.7	7.1	34	70.2
5 m	4.2	8.0	34	71.1
6 m	4.6	8.7	34	71.8
7 m	4.9	9.4	34	72.5
8 m	5.3	10.1	34	73.2
9 m	5.6	10.7	34	73.8
10 m	5.9	11.3	34	74.4
100 m	18.6	35.7	34	98.8
500 m	41.7	79.9	32	139.2
1 km	58.9	112.9	31	170.4
2 km	83.3	159.7	28	211.7
3 km	102.0	195.6	25	242.0
4 km	117.8	225.8	22	266.6
5 km	131.7	252.5	20	289.6
6 km	144.2	276.5	18	309.9
7 km	155.7	298.7	16	328.4
8 km	166.4	319.3	15	347.1
9 km	176.5	338.6	14	364.6
10 km	186.0	356.9	12	379.2
15 km	227.5	437.0	7	450.0
50 km	412.0	796.0	0	796.0
100 km	575.9	1122.1	0	1122.1
200 km	796.5	1576.8	0	1576.8
300 km	955.0	1919.0	0	1919.0
300 km	955.0	1919.0	0	1919.0
400 km	1080.5	2202.1	0	2202.1
500 km	1184.6	2446.9	0	2446.9

their orbiting Command Module. The landing gear remains there to this day, surrounded by the excess equipment that the astronauts abandoned, though it was worth a million dollars at the time. The seismometer, solar-wind detector, and American flag still stand exactly as they were.

Subsequent NASA missions contributed their share of astrotrash, too. When Alan Shepard was practicing his golf swing on the Moon, he whiffed at the first ball and shanked a second shot. He drove a second ball into a distant bunker. He left a ball behind "for future golfers." The early pioneers got away with a lot. Nowadays, such a flagrant violation of the Lunar Litter Law of 2018 would be punishable by a large fine and 80 hours of community service at a garbage recycling center. Nothing is wasted on the Moon, or the Earth for that matter, these days.

JACK AND JILL

MUNDILFARI HAD A WONDERFUL BOY NAMED MOON AND A BEAUTIFUL DAUGHTER called Sun—so goes an old Norse tale. These names angered the gods, who decided to take the children up to heaven, where the daughter became the coachdriver to the Sun and the son the coachdriver to the Moon.

In another version of this tale, the Moon goddess took pity on two other children, Hjuki and Bil, whose father had ordered them day after day to carry water from a well. According to this myth, she brought them to live with her and the dark spots on the Moon are the children Hjuki and Bil.

You can hear an echo of this legend in the well-known English nursery rhyme:

> *Jack and Jill went up the hill*
> *To fetch a pail of water.*
> *Jack fell down and broke his crown*
> *And Jill came tumbling after.*

The names Hjuki and Bil are also related to the Moon. The verb *jukka* means "to grow," while the verb *bil* means "to fade away." These clearly refer to the waxing and waning of the Moon.

The pail of water is said to symbolize the Moon's influence over the waters of the Earth and rainfall.

HISTORY AND PEOPLE

Is there life on the Moon? Are there any signs of intelligent life in the crevasses dotting the craters and plains? Ever since they first cast eyes on it, humans have imagined that the Moon must be inhabited. Though it wasn't, this fact was far from certain until the first manned landing on the Moon.

The Moon Museum's display (see "Moon Museum" in the Out and About chapter) includes a few striking examples of how humankind first pictured life on the Moon and our attempts to reach it. And it documents our successes.

VISIONARIES AND DREAMERS

Lucian of Samosata
In the 2nd century A.D., the Greek rhetorician Lucian of Samosata wrote a satirical account of a journey to the Moon. According to Lucian, the Moon was inhabited by the souls of human beings, for after we die we either wind up on the Moon in a kind of paradise or else in caves and cracks. Sad to say, Lucian failed to provide any details about his spaceship.

Kepler's Dream
The first people to observe the Moon were absolutely convinced that there was life on the Moon. One of the first Moon-watchers was Johannes Kepler (1571-1630), a German astronomer who formulated the laws of planetary motion. Kepler wrote a remarkable book called *Somnium* ("The Dream"), which was published posthumously in 1634. In this science-fiction fantasy, Kepler described an imaginary journey to the Moon and back, noting everything he expected to find there.

Kepler's story began with the discovery of a bridge between the Earth and the Moon. This bridge was built of a shadow, the shadow cast by the Earth into the celestial void. It wasn't the strongest of bridges, but it was enough to accommodate Kepler's travelers. When the top of the shadow reached the Moon (which happened every time there was a lunar eclipse), it formed a link between the Earth and the Moon. All you had to do to reach the Moon was to glide over this shadow. When the Moon cast its

A Jules Verne type of fantasy: a Moon train. Note that the compartments are divided into first, second, and third class.

shadow on the Earth (which occurred every time there was a solar eclipse), you could make the return trip to Earth. There was only one catch: you had to be as light as a feather to walk on the shadow.

After reaching the Moon, the travelers found themselves in a dark world. Dark hollows and dusky entrances to caves inhabited by Moon creatures rose up before them. The caves appeared to be fortresses that the Moon creatures had built for protection against the scorching heat of the Sun during the day and the icy cold at night.

The Moon creatures glided eerily under the foliage of exotic trees. They were snakelike monsters weighing more than humans. Many of them had wings, as if they were a cross between a bird and a snake.

HISTORY AND PEOPLE 43

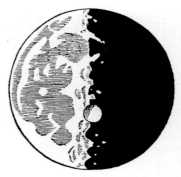

In 1609 Galileo Galilei was the first to aim a telescope at the Moon. He described his celestial discoveries in his book Sidereus Nuncius (The Starry Messenger), *which includes this drawing of the Moon as he saw it through his telescope.*

Galileo Galilei (1564–1642)

Others kept to the water. Very few of them could endure the Sun's intense heat. Most of them simply rested during the long lunar day and came to life only at night. When they looked into the sky, one celestial object was brighter and more luminous than all the rest. This was Earth, which in their language was known as "Volva." Only one side of Volva was visible; the other side was lost to them forever.

At the time Kepler was writing his book, a pair of Dutch glass-grinders named Zacharias Jansen and Hans Lippershey were busy making the first telescopes. Soon the Italian astronomer and physicist Galileo Galilei (1564-1642) was aiming one of these newfangled devices at the heavens, making him the first person on Earth to observe the craters on the Moon.

It didn't take Kepler long to acquire his own telescope, so that he too could admire the Moon and his mythical caves. Yet his telescopic observations only served to confirm to him that the Moon's craters had been built by Moon creatures.

The Man in the Moon

In the beginning of the 17th century, the English bishop Francis Goodwin wrote his exotic tale, *The Man In The Moon: Or A Discourse Of A Voyage Thither*, about a Spanish sailor named Domingo Gonsales. After Gonsales was abandoned on an uninhabited island, he was determined to get back home. Step one in his rescue plan was to train some wild swans to carry

> ## FATHER MOON
>
> THE MOON WAS RULER OF THE EARTH, ACCORDING TO THE SIRIONO OF eastern Bolivia. Moon created all life on Earth. This is their story of why the Moon went up into the night sky:
>
> > *Yasi (Moon) had a son. One day a jaguar was playing with the boy and accidentally killed him. Yasi wanted to find out who the killer was, but none of the animals would tell him. So the enraged Yasi gave the howler monkey a long neck, put spines on the porcupine, and made the tortoise slow.*
> >
> > *Still infuriated, Yasi leapt up into the sky, where he spends half his time hunting. When he returns, his face dirty from the hunt, he washes a little each day until the Moon is full again. When he leaves, his face gets dirtier each day until it disappears from view.*

heavy weights. Once they'd managed that, Gonsales added a chair. By way of experiment, he tied a lamb to the chair, and the swans flew the lamb over the island. In the hope that the swans would fly him back to Spain, Gonsales climbed into the chair, and the swans carried him aloft. Unfortunately, they got out of control and flew him to the Moon.

Cyrano de Bergerac
A few years later the French poet Cyrano de Bergerac also described a trip to the Moon. Cyrano assumed that when day dawned, dew was sucked upward by the Sun. He reasoned that a man wearing vials full of dew would likewise be drawn toward the Sun. So he collected some dew in vials, tied them around his waist, and launched himself by the power of evaporation. He rose so fast that he had to break a couple of bottles to keep from whizzing past the Moon. However, he miscalculated, broke too many, and slipped back into the Earth's gravitational field.

Undaunted, Cyrano planned a second attempt, this time using a flying chariot. One day some soldiers attached firecrackers to his spaceship, and Cyrano came along just as they were lighting the fuses. He leapt aboard to put out the fire but he was too late. He was propelled to the Moon in a blaze of fireworks. To his surprise the Moon was inhabited. He even ran into a familiar figure: Domingo Gonsales, the Spanish sailor from the Goodwin story.

Gruithuisen's Lunar City

Improvement in telescopes and the gradual expansion of human knowledge about the Moon had little effect on the conviction that its gigantic structures were the work of intelligent forms of life.

In 1822, Franz von Paula Gruithuisen (1774-1852), a German health officer and zealous Moon observer, saw what he believed to be an entire city on the Moon. In the partially collapsed Schröter Crater, Gruithuisen thought he could make out a fortified structure with regularly formed walls. He theorized that the inhabitants of the Moon, whom he called "Selenites," had built this citadel for defense. A star-shaped structure near the citadel "proved" that the Selenites worshipped the stars. Gruithuisen filled 25 thick logbooks with his notes.

In the 17th century Cyrano de Bergerac thought he could reach the Moon by tying bottles of dew to his belt.

> *In all probability, we have before us a city with underground vaults, where the Selenites are forced to live to protect them from the intense heat of the day and the cold at night. Only the upper side is visible. It is apparently covered with soil, because everything is of the same dark shade. The two straight dikes must have been constructed to block the cold passing wind that rages through the southwestern mountains. Only the eastern part of the citadel is less regular; perhaps this is merely a newer addition to the city. The lunar architects must have achieved a certain degree of sophistication, because the structure is oriented according to the directions of the wind.*
>
> *In addition, the 'spindle' on the top is a well-sheltered municipal park, where the Selenites can come to catch their breath from time to time. What about religion, you ask? See the lower right. There you can make out a star-shaped configuration resembling a bulwark which is linked to the other structures. It might be a kind of temple, dedicated to the star-gazers. Don't forget that on the Moon the stars can also be seen during the daytime.*

Only later did the German astronomer Johann Heinrich von Mädler, head of the observatory in Dorpat, prove that Gruithuisen's city was nothing more than a relatively low mountain range. An impact crater at one end had left an impression resembling a citadel.

SEEING DOUBLE

THE FIRST STEREO PHOTOGRAPHS OF A PLANETARY BODY CREATED A SENSATION at the International Exhibition of 1862 in London. Observers could view the striking stereo pairs of the full Moon through a stereopticon, a 3-D viewing device that was extremely popular during the late 19th and early 20th centuries.

English astronomer Warren de la Rue, who took the photos at the Yerkes Observatory, was able to make a stereo image of the Moon from Earth because of librations: motions of the Moon and Earth that enable an observer on Earth to see the lunar surface from slightly different angles at different times.

the city Franz von Paula Gruithuisen thought he saw in Schröter Crater

Herschel's Unicorn

Even after that, observers continued to let their imaginations run wild. The British-born essayist Richard Adams Locke (1800-71) even managed to fool the scientists. It all began on 25 August 1835, with the publication of an article in *The Sun,* a New York daily. Under the sensation-grabbing headline "Great Astronomical Discoveries Lately Made by Sir John Herschel at the Cape of Good Hope," *The Sun* reported that the English astronomer Herschel, using a 7,000 kg telescope lens with a magnification of 42,000, had been able to observe rocks, trees, flowers, and intelligent winged beings of both sexes on the Moon.

All of New York was in a frenzy, with everyone naturally dying to know more about the creatures on the Moon. In the first installment, Locke relied heavily on the power of suggestion. First he gave Herschel's credentials as an astronomer (his father discovered Uranus in 1781) an enormous buildup. Then he announced that the newspaper owed this exclusive article on Herschel's latest discoveries to his faithful assistant of many years, Dr. Andrew Grant. According to the article, Herschel had given the data to Grant to pass on to the Royal Society, the most respected scientific institute in all of England. Thanks to a generous gesture by Dr. Grant, the readers of the *Edinburgh Journal of Science* were able to share in the discovery. And luckily *The Sun* had been able to obtain an advance copy of the journal so that its readers were the first to hear this exciting news.

The readers of *The Sun* got their money's worth. It unveiled one spectacular discovery after another. The second installment was written as if the Moon had come down to the Earth, and the readers could stroll around on it with the observers. Gazing through their telescope, Herschel and Grant began their "walk" on a lunar landscape dotted with basalt-like rocks. Suddenly they saw a prominent cliff. It was covered with crimson flowers. "Precisely similar," remarked Dr. Grant, "to the *Papaver rhoeas*, or rose-poppy of our sublunary cornfield." For the first time ever, one of nature's organic products had manifested itself to mankind in a totally alien world.

But poppies were not all the Moon had to offer. At the foot of another rocky mass they could see an entire lunar forest. "The trees," reported Dr. Grant, "for a period of ten minutes, were of one unvaried kind, and unlike any I have seen, except the largest class of yews in the English church-yards, which they in some respects resemble."

Next they noticed a green plain on which some kind of fir tree abounded. They aimed the telescope at a different spot, and their eyes fell on an alpine scene, complete with a lake. The marine-blue water broke in large white billows upon the shore . . .

After they had been gazing at these wonders for two hours, Herschel proposed that they move on and subject other important valleys to a thorough inspection. All they had seen before paled in comparison to the panoramas which then unfolded before their eyes. Irregular groups of crystals, like wine-colored amethysts, rose from the ground in obelisks and pyramids. Lying behind the crystals was a plain filled with herds of tiny bison. A shaggy pelt extended across their faces, presumably to protect their eyes from the extremes of light and darkness on the Moon. Then blue goatlike creatures appeared. The males had beards and only one horn. The females had neither a beard nor a horn, but did have a longer tail.

(above) During a total solar eclipse, the surface of the Moon is bathed in a crimson glow. The simultaneous sunrise and sunset surrounds the Earth with a reddish-orange ring. *(Erven J.J. Tijl Press, Zwolle, Netherlands)* *(below)* Apollo 11 in the Sea of Tranquility *(July 1969)*. The Moon Museum was built at this site in 2015. *(NASA)*

A batman, a lunar antelope, and a lunar bison. According to Locke's Moon hoax, these creatures were observed by the famous British astronomer Sir John Herschel. (Lee Battaglia)

HISTORY AND PEOPLE 49

Richard Adams Locke astounded the world in 1835 with his story of how living creatures were discovered on the Moon. Only later did it prove to be a hoax.

The second installment ended with this description of the unicorn. New Yorkers fought over *The Sun*, whose circulation had risen from 8,000 to a dizzying 19,360 practically overnight. Suddenly *The Sun* had become the world's largest newspaper. Even the London *Times* had a circulation of only 17,000.

The following installment reported on the discovery of intelligent life.

> *We were thrilled with astonishment to perceive four successive flocks of large winged creatures . . . descend with a slow even motion from the cliffs on the western side, and alight upon the plain. . . . We counted three parties of these creatures, . . . walking erect towards a small wood near the base of the eastern precipices. Certainly they were like human beings, for their wings had now disappeared, and their attitude in walking was both erect and dignified. . . . [T]hese creatures were evidently engaged in conversation; their*

gesticulation, more particularly the varied action of their hands and arms, appeared impassioned and emphatic. We hence inferred that they were rational beings.

In the last installment, Locke promised that Sir John himself would comment at length on his discoveries. But before that article was due to appear, *The Sun* prepared a pamphlet containing the entire story. The deception was near complete. Even mathematicians did not doubt the authenticity of the stories.

Still, not everyone could be fooled so easily. Two professors from Yale University presented themselves at *The Sun* and asked for permission to peruse the mathematical data that the editors had omitted in the article. Locke told them that the original was at the printer's and gave them the address. The two scholars raced off in pursuit, but Locke was faster. He instructed the printer to send the gentlemen to another address. In the end, the professors gave up. They suspected they were victims of a trick but couldn't prove it.

It wasn't until *The Journal of Commerce,* another New York paper, wanted to reprint the entire story that the hoax came to light. At first

Locke's story in The Sun *did not contain any illustrations, though they were added when it was reprinted in several different languages. This drawing of "bat people" on the Moon appeared in an English pamphlet published in 1836.*

Locke tried to dissuade the editors from printing it by pointing out that it was stale news. But he was finally forced to tell the truth. And so *The Journal of Commerce* became the first paper to announce that the whole story had been a fabrication, from beginning to end.

And what about Sir John? He was totally in the dark. He saw *The Sun* and the pamphlet edition for the first time when an American visitor came to see him in Cape Town. Dumbfounded, he inquired if that drivel had actually appeared in the *Edinburgh Journal of Science*. His visitor assured him it had not—the journal had gone out of business years before— and that the whole thing had been a hoax. After that, Herschel was able to laugh at the joke. But in the meantime the story had spread to Europe, where many astronomers failed to appreciate the humor. Years passed before they received findings from American astronomers without suspicion.

Bugs and Bombs

In the beginning of the 20th century, the famous American astronomer William Henry Pickering (1858-1938) reported observing dots moving eerily at the bottom of a lunar crater. These dots changed constantly in shape and size. Pickering believed them to be the shadows of swarms of flying insects, which, he surmised, lived in the crater because it contained some moisture. He conjectured that when the rising sun began to warm the lunar soil, the insects crept out of their eggs and began to forage for air and water in the crater. When night descended on the Moon, the bugs froze into a kind of suspended animation until the Sun's first rays brought an end to the icy night and roused them to life. Pickering also believed that the different colors he observed in certain crater floors were signs of a primitive form of plant growth. Moss and alga-like plants supposedly bloomed briefly during the day until the nighttime chill ended their short lives.

In 1951, soon after the beginning of the atomic age, a Spanish engineer named Sixto Ocampo proposed a nuclear-war theory to explain lunar geography. Ocampo postulated that the craters on the Moon were the result of an atomic war between two Moon tribes. That some of the craters had central mountain peaks and others did not was for him sufficient evidence for the existence of two armies using two different types of bombs.

At the time of the first manned landing on the Moon (1969), scientists were still uncertain whether moondust might harbor bacteria or some other microscopic life-form. When the crew returned to Earth, splashing down in the Pacific Ocean, Navy frogmen opened the hatch of the space capsule and tossed insulated suits to the three astronauts. If the astronauts had brought back living organisms from the Moon, the suits would have prevented them from spreading.

That was only the first of a series of complicated measures designed to rule out any risk of contamination. The astronauts were also not welcomed as lavishly as their predecessors had been. Okay, so the president of the United States was on board a nearby aircraft carrier. But he was al-

---

ANIMALS OF THE MOON

MANY ANCIENT CULTURES LINKED ANIMALS WITH THE MOON. IN MEXICAN mythology, Mayauel was the goddess of the night sky; she nourished the stars, who were thought of as the fish of the heavens. She sat upon the tortoise's head—he was the guardian of the Moon and retreated into the shadows with the Moon.

Ancient Egyptians associated cats with the Moon. The cat goddess represented the Moon's power over pregnant women. Her son was believed to be the Moon itself; his light would make women fruitful, so that the human seed might grow in the mother's womb.

Eskimos thought the Moon was home to Sedna, the Old Woman who was the keeper of all mammals. When offended by humans, she could seize the lakes, rivers, and oceans of the world.

Animal stories related to the Moon frequently contain an element of foreboding, especially with regard to the wolf. It is said that when the Moon vanishes completely, this will mark the end of the world. According to an apocalyptic Norse legend, one day the wolf will eat the Moon, and then the heavens will be splattered with human blood, the Sun will dim, and a great wind will stir. Then will come the eternal winter, when war will break out, the sea will bubble, and the earth will tremble.

Old German legends tell of the wolves Sköll and Feunir, who tried to swallow the Sun and the Moon. Humans watching the spectacle raised such an outcry that the wolves dropped their prey at the very last minute. Even today, many tribal peoples still make a racket during solar and lunar eclipses.

lowed to view the events only from afar. The astronauts themselves were immediately transferred to an isolation container, which was to remain their quarters until the end of the sea voyage. The space capsule was hauled on board and linked to the container by a plastic tunnel. The astronauts crawled back into the space capsule one last time to retrieve the lunar soil samples and film rolls. After that, the container was hermetically sealed. The astronauts remained in quarantine for 21 days, until it had been proved without a doubt that they were not contaminated.

FROM FIRECRACKER TO ROCKET

Wan Hu
In all likelihood, humankind's first spaceship was made in China. Around the year 3000 B.C., Mandarin Wan Hu set about to build a spaceship that, according to his design specifications, was to be propelled by 47 firecrackers. Two kites attached to a rope were added to help maintain balance. Wan Hu's servants lit the firecrackers, and the mandarin took off in a blaze of fire. History neglects to tell us what became of the world's first astronaut, but we can only assume the worst.

Tsiolkovsky
The longing to fly through the air like a bird and travel to the Moon has long been the wish of many an Earth-bound creature. Still, it wasn't until the beginning of the 20th century that the Russian Konstantin Eduardovich Tsiolkovsky (1857-1935) calculated the amount of thrust needed to propel a rocket into space.

One of Tsiolkovsky's proposals was to use liquid fuel in the rockets. From firecracker to rocket—it was a giant leap.

MODERN-DAY TRAVELERS

The Space Race
The Moon Museum exhibit also documents the early years of space exploration. It begins with the first launching of a rocket on 16 March 1926 by the American scientist Dr. Robert Hutchings Goddard, moves on to the

horribly destructive v-2 guided missiles fired by the Germans during WW II, and details the space race between the Russians and the Americans. President John F. Kennedy started the contest on 25 May 1961, when he announced America's intention to land a man on the Moon and return him safely to Earth before 1970. On 20 July 1969, after spending billions of dollars on the Apollo program, the Americans did indeed beat the Russians to the Moon.

Russian Successes
The race to the Moon was exceptionally exciting. On 2 January 1959 the Russians launched Luna 1, a spacecraft that flew by the Moon at a distance of 5,955 km. On 12 September 1959 they launched Luna 2, which crash-landed on the Moon three days later.

RUSSIAN UNMANNED MISSIONS TO THE MOON

NAME	LAUNCH DATE	LANDING SITE (LAT./LONG.)	REMARKS
LUNA 1	2 JAN. 1959	—	FLY-BY AT 5,955 KM
LUNA 2	12 SEPT. 1959	30°N/0°E	CRASH LANDING; FIRST LUNAR IMPACT
LUNA 3	4 OCT. 1959	—	FIRST FAR-SIDE PICTURES (70% OF LUNAR FAR SIDE PHOTOGRAPHED)
LUNA 4	2 APRIL 1963	—	SOFT LANDING ATTEMPT; FLY-BY AT 8,529 KM
LUNA 5	9 MAY 1965	1° 42'S/25°W	SOFT LANDING ATTEMPT; CRASHED
LUNA 6	8 JUNE 1965	—	SOFT LANDING ATTEMPT; FLY-BY AT 161,000 KM
ZOND 3	18 JULY 1965	—	RETURNED PICTURES OF REMAINING PART OF LUNAR FAR SIDE
LUNA 7	4 OCT. 1965	9° 48'N/48°W	SOFT LANDING ATTEMPT; CRASHED
LUNA 8	3 DEC. 1965	9° 08'N/63° 18'W	SOFT LANDING ATTEMPT; CRASHED
LUNA 9	31 JAN. 1966	7° 08'N/64° 33'W	FIRST SUCCESSFUL SOFT LANDING; RETURNED TV PANORAMAS AND RADIATION DATA

Name	Launch Date	Landing Site (lat./long.)	Remarks
Luna 10	31 March 1966	—	First successful lunar orbit
Luna 11	24 Aug. 1966	—	Lunar orbit
Luna 12	22 Oct. 1966	—	Lunar orbit; returned TV pictures of surface
Luna 13	21 Dec. 1966	18° 52'N/62° 03'W	Soft landing; tested soil
Luna 14	7 April 1968	—	Lunar orbit
Zond 5	14 Sept. 1968	—	Circumlunar fly-by
Zond 6	10 Nov. 1968	—	Circumlunar fly-by
Luna 15	13 July 1969	Mare Crisium	Soil sample return attempt; crashed 21 July 1969 during Apollo 11 mission
Zond 7	7 Aug. 1969	—	Circumlunar fly-by returned color pictures
Luna 16	12 Sept. 1970	0° 41'N/56° 18'E	First successful automatic lunar soil sample return; 100 g of soil returned to Earth
Zond 8	20 Oct. 1970	—	Circumlunar fly-by
Luna 17	10 Nov. 1970	38° 17'N/35°W	Lunokhod 1: first successful rover; French laser reflector
Luna 18	2 Sept. 1971	3° 34'N/56° 30'E	Crashed
Luna 19	28 Sept. 1971	—	Lunar orbit
Luna 20	14 Feb. 1972	3° 32'N/56° 33'E	Automatic soil sample return
Luna 21	8 Jan. 1973	27°N/31°E	Lunokhod 2
Luna 22	29 May 1974	13°N/62°E	Lunar orbit
Luna 24	9 Aug. 1976	12° 45'N/62° 12'E	Deep soil sample return (1.78 meters)

RUSSIAN LUNAR ROVERS

Name	Weight (kg)	Distance Traveled (km)	Remarks
Lunokhod 1	756	10.5	Luna 17. Mare Imbrium. Operated more than 11 months after landing on 11-17-70. Returned more than 200 TV panoramas and 20,000 pictures
Lunokhod 2	850	37	Luna 21. Le Monnier. Operated four months after landing on 16 January 1973. Returned 86 TV panoramas and 80,000 pictures

On 12 September 1959, the unmanned Luna 2 crashed on the Moon. The first human-made object to reach the Moon left behind these insignias as a lasting memory of Russian achievement.

THE SOVIET MOUNTAINS— A LUNAR GLITCH

THE RUSSIANS MADE HISTORY IN 1959 WHEN THEY SENT THE FIRST LUNAR PROBE past the far side of the Moon. Luna 3 photographed a clear stripe, which the Russians proudly dubbed the Soviet Mountains (Montes Sovietici). But American photographs taken later made it clear that the Soviet Mountains were nothing more than a transmission error. Until 1978, the Russians refused to recognize the error and continued to print lunar atlases with the Soviet Mountains.

The subject was broached once more during the 17th General Meeting of the International Astronomical Union in 1979. By that time the Soviet Mountains had also disappeared from the Russian atlases. Somebody in Moscow had apparently decided that, faced with the American astronomers and the scads of far-side photographs taken during the Apollo program, the Soviet Mountains were a lost cause. Still, the Russians had no desire to go home empty-handed, so they submitted a list of eight new names for craters on the far side of the Moon. One of these, Lipsky, was drawn as a nice, round crater on the Russian maps. But the American photos showed nothing in that spot. To keep the matter from escalating further, the U.S. and Russia decided after mutual deliberation to call the area the Lipsky Plain.

It was the first human-made craft to set down—albeit in pieces—on the Moon. In the 1970s the Russians also managed to collect lunar soil samples by using robot vehicles. In 1970 and 1973 they landed remote-controlled roving vehicles called "Lunokhods" on the Moon. Guided from Earth, these unmanned lunar rovers explored the lunar surface. You can still find some of these early Russian rovers at their original landing sites. (See "Astrotrash" under "Environmental Issues.") For those who wish to view them, the spacecraft the Russians sent to the Moon are listed in the charts.

American Space Exploration

The Americans got off to a rockier start. They initially trailed far behind, but wasted no time in showing that they could do everything the Russians could do. The first manned flights of the Mercury program were followed by the Gemini program. Then, to prepare for a Moon landing, the Americans developed three unmanned programs: the Ranger program, the Surveyor program and the Lunar Orbiter program.

The Ranger Program

In the Ranger program, unmanned lunar probes equipped with cameras crashed into the lunar surface, snapping pictures all the way. In 2017, parts of what are assumed to be Ranger 8 were found not far from the Moon Museum.

Souvenir hunters can have a field day on the Moon. The rule regarding the remains of hard landings is finder's keepers. However, all findings must be registered to aid later research. For those wanting to try their luck, the relevant data and coordinates of the impact sites of the Ranger flights are listed in the chart.

The Surveyor Program

The Surveyor program concentrated on landing automatic three-legged robots on the lunar surface. About 35 minutes before landing, the Surveyors were turned so that their legs were aimed downward. At about 100 km above the lunar surface the main retro-rocket began to fire. At an altitude of eight km the speed had been reduced from 10,000 km per hour to 400 km per hour; about here the burned-up main retro-rocket was cast off and the smaller retro-rockets took over. Traveling at five km per hour, the Surveyors stopped at four meters above the ground and dropped to the Moon in free fall. When they finally plopped down on the lunar surface, they were traveling 16 km per hour, a speed the landing gear had been designed to accommodate.

THE RANGER PROGRAM

No.	Launch Date	Impact Site (Lat./Long.)	No. of Pictures	Remarks
1	23 Aug. 1961	—	—	High Earth orbit attempt; one week in too low an orbit
2	18 Nov. 1961	—	—	High Earth orbit attempt; one day in too low an orbit
3	26 Jan. 1962	—	—	Lunar fly-by at 36,808 km
4	23 April 1962	15° 30'S/130° 42'W	—	Crashed on lunar far side
5	18 Oct. 1962	—	—	Lunar fly-by at 735 km
6	30 Jan. 1964	9° 20'N/21° 31'E	—	Crashed; no pictures returned
7	31 July 1964*	10° 36'S/20° 36'W	4,308	Returned 4,306 pictures before impact on Mare Cognitum
8	20 Feb. 1965*	2° 36'N/24° 48'E	7,137	Returned 7,137 pictures before impact on Mare Tranquilitatis
9	24 March 1965*	12° 48'S/2° 24'W	5,814	Returned 5,814 pictures before impact in Alphonsus Crater

* Impact date

THE SURVEYOR PROGRAM

No.	Date of Landing	Landing Site (Lat./Long.)	No. of Pictures
1	2 June 1966	2° 32'S/43° 19'W	11,237
2	20 Sept. 1966*	4°N/11°W	—
3	20 April 1967	3° 11'S/23° 23'W	6,315
4	14 July 1967*	0° 26'N/1° 30'W	—
5	11 Sept. 1967	1° 25'N/23° 11'E	19,054
6	10 Nov. 1967	0° 32'N/1° 24'W	30,396
7	10 Jan. 1968	40° 52'S/11° 28'W	21,091

* Launch date

The unmanned Surveyors didn't quite land according to plan. Surveyor 3 bounced up and down a few times before coming to a standstill, all the while photographing the imprint of its footpads and relaying the pictures back to Earth.

Surveyor 3 landed on a slope and bounced up and down a few times, enabling the craft to photograph its own footprints. These photographs proved that the Moon was not covered with an inordinately thick layer of dust, which was good news, since it meant that human beings would be able to walk on the Moon. Later Surveyors dug trenches in the lunar soil and analyzed the soil samples.

The Lunar Orbiter Program

The Lunar Orbiter program systematically mapped the Moon from outer space. The primary objective of the program was to select potential landing sites for future Apollo missions. Each Lunar Orbiter had two cameras on board: one with a wide-angle lens and one with a telephoto lens. After their photographs were relayed back to Earth, the Lunar Orbiters were deliberately crash-landed on the Moon.

THE LUNAR ORBITER PROGRAM

No.	Launch Date	Impact Date	Impact Site (Lat./Long.)	No. of Pictures/ Wide Angle
1	10 Aug. 1966	29 Oct. 1966	6° 21'N/160° 42'E	206/29
2	6 Nov. 1966	11 Oct. 1967	3° N/119° 06'E	207/616
3	5 Feb. 1967	9 Oct. 1967	14° 18'N/92° 42'W	164/481
4	4 May 1967	6 Oct. 1967	22° 30'W	128/447
5	1 Aug. 1967	31 Jan. 1968	2° 48'S/83° 02'W	211/628

APOLLO MOON LANDINGS

Apollo Mission No.	Landing Date	Time Spent on the Moon	Distance Traveled	Soil Samples (kg)	Landing Site (Lat./Long.)
11	20 July 1969	21:36	250.0 m	20.7	0° 41'N/23° 26'E
12	19 Nov. 1969	31:31	2.0 km	34.1	3° 11'S/23° 23'W
14	5 Feb. 1971	33:30	3.3 km	42.8	3° 40'S/17° 28'W
15	30 July 1971	66:55	27.9 km	76.6	26° 06'N/3° 39'E
16	21 April 1972	71:00	27.0 km	96.7	8° 59'S/15° 31'E
17	11 Dec. 1972	75:00	35.8 km	15.0	20° 10'N/30° 45'E

The Apollo Program

A culmination of the foregoing programs, the Apollo program was geared toward sending manned missions to the Moon. After a few initial test flights, the U.S. in December 1968 launched Apollo 8, the first manned mission to orbit the Moon.

Apollo 10 (May 1969) was a dress rehearsal for the upcoming Moon landing, though the spacecraft remained at least 15 km above the lunar surface.

Apollo 11, launched on 16 July 1969, finally landed the first two human beings on the Moon: Neil Armstrong and Edwin Aldrin. The third astronaut, Michael Collins, remained in the Command Module, which continued to orbit the Moon. Three craters named after these pioneers lie near the Moon Museum.

WORDS—
SETTING THE RECORD STRAIGHT

OFFICIALLY, THE FIRST WORDS SPOKEN ON THE MOON WERE "Houston, Tranquility Base here. The Eagle has landed." But before Neil Armstrong spoke these words, his co-pilot Edwin Aldrin already had murmured, "Contact light. Okay, engine stop."

Armstrong descended the ladder and announced: "That's one small step for man, one giant leap for mankind." But man and mankind in this instance means the same thing. The article "a" was later inserted into the record to amend the message to "one small step for a man," referring to the short gap between the ladder and the ground that required Armstrong to jump down. By some accounts, Armstrong says he did say the "a"—it just didn't get transmitted properly.

Officially, the last words spoken on the Moon were those of Eugene (Gene) Cernan of the Apollo 17 mission on 14 December 1972, 5:41 a.m. GMT: "I take man's last steps from the surface for some time to come, but we believe not too long into the future," he said as he paused before climbing his LM's ladder. "We leave the Moon as we came, and, God willing, as we shall return, with peace and hope for all mankind. Godspeed from the crew of Apollo 17!" But when he closed the door he said just before lift-off: "Let's get this mother out of here!"

The Lunar Lander of the Apollo 11 mission was named the Eagle. On 20 July 1969, at 8:17 p.m. GMT, the first words to be spoken from the Moon crackled over the radio to Mission Control: "Houston, Tranquility Base here. The Eagle has landed." Six hours and 38 minutes later, Neil Armstrong clambered down the Eagle's ladder and uttered the historic words, "That's one small step for a man, one giant leap for mankind," before planting his foot on the Moon.

Astronauts carried out a variety of experiments during the Apollo missions. They placed laser reflectors, seismometers, magnetometers, and solar-wind collectors on the Moon. They gathered 385.9 kg of soil samples. A Lunar Roving Vehicle sent along on the last three missions enabled the astronauts to cover much larger distances.

Banzai!

On 10 April 1993, Japan entered the Moon race when the Japanese Institute of Space and Astronautical Science (ISAS) decided to crash-land its Hiten satellite (named after a goddess of music) on the Moon. Hiten was put into orbit around the Moon in February 1992 after an earlier attempt with its smaller subsatellite Hagoromo failed. An astronomer at the Anglo-Australian Observatory in Australia observed the impact of the kamikaze satellite. The observatory's infrared imaging spectrometer recorded half a dozen frames as the flash lit up the lunar night. The intensity of the burst and the apparent lack of a dust cloud indicated that Hiten had hit solid rock, converting nearly all its impact energy to heat and light.

THE LUNAR AGE

CLEMENTINE AND THE SOUTH POLE

Lunar exploration was first encouraged by the political imperatives of the Cold War that marked the 1960s and early 1970s. But after the last lunar landing in 1972 (Apollo 17), exploration virtually came to a halt for two decades.

A revival of lunar exploration came unexpectedly with the Clementine mission to the Moon. The Clementine, an unmanned spacecraft that orbited the Moon for 71 days in 1994, was a military mission that piggybacked astronomical experiments. Clementine recorded about 1.5 million images in 11 visible and near-infrared colors. A mosaic of 1,500 images of the lunar South Pole was particularly intriguing. It revealed for the first time a depression four billion years old and 2,500 km wide near the pole.

The depression is christened the South Pole-Aitken Basin. Within the basin is an impact crater 12 km deep—more than seven times as deep as the Grand Canyon—and 300 km wide. It's by far the deepest impact crater in the solar system.

A quarter of the world's population was watching television on 21 July 1969 as Neil Armstrong walked on the Moon. (NASA)

Part of the basin stretches to the South Pole itself. There, at Amundsen crater and the Mountain of Eternal Light, sunlight never penetrates. Since it remains at a frigid -230° C, it became a perfect icy storehouse for water that comets brought to the Moon.

"This is the place on the Moon where you would go to get ice for your cocktail," joked Clementine investigator and geologist Eugene M. Shoemaker when he saw the photographs.

Later missions did find ice deposits, though they were too dirty for immediate use in cocktails. But after purification the ice proved to be a most valuable source of drinking water. The ice's oxygen and hydrogen molecules are also used in the production of spacecraft fuel.

THE DISCOVERY OF ICE

In 1997, the world saw the tiny impact crater that was all that was left of Hiten (see "Banzai," above) when the Japanese lunar orbiter Lunar-A visited the grave site and relayed the images to Earth.

LUNAR ARCHIVES

EARTH'S ATMOSPHERE MAKES IT VERY DIFFICULT TO KEEP ARCHIVES. PRINTED materials and film are slowly corroded by the acidity of the air, which eventually also destroys magnetic tapes and even CD-ROMs.

That makes the lunar South Pole, with its absence of air, scarce illumination, and deep-freeze temperatures, an ideal site for the Lunar Archives. The archives, to be completed in 2026, will include the collections of all major libraries of Earth. They will be stored underground to protect books and tapes from meteorites and cosmic radiation, the only "weathering" agents that still exist on the lunar surface.

An interplanetary archive was proposed as far back as 1994 by astronomer Jacques Beckers of the European Southern Observatory headquarters in Garching, Germany. He envisioned a library that could not be destroyed by environmental effects, earthquakes, floods, or wars, and that could serve as the ultimate repository of human knowledge. Even aliens visiting our solar system could consult the archives long after the human race has disappeared.

*(above) The central peak of Alphonsus Crater sometimes displays slight volcanic activity.
(below) Epsilon peak near the Moon's South Pole, also known as the Mountain of Eternal Light
(Illustrations: Erven J.J. Tijl Press, Zwolle, Netherlands)*

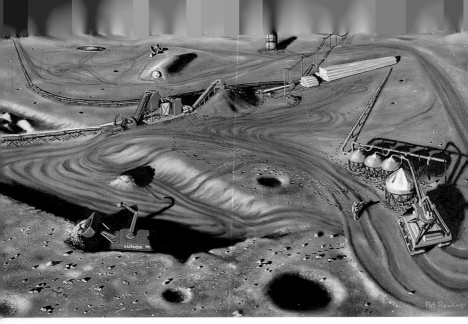

(above) The Lunox company's oxygen and concrete production operation is excavating near Mt. Schneckenberg. *(Pat Rawlings/NASA)* *(below)* Mont Blanc Resort is the most luxurious of the lunar bases. In the middle it boasts a park where trees and plants are grown under a dome made out of glass extracted from Moon rocks. *(Jinsei Choh/Ohbayashi Corporation)*

Lunar-A marked the beginning of the Lunar Age, since it found huge amounts of ice at the ever-dark crater at the South Pole. A few months earlier, the NASA satellite Lunar Prospector had discovered what scientists suspected to be ice, but confirmation had to wait until one of Lunar-A's probes was shot into the lunar surface. The probe, which struck the Moon at 300 meters per second, penetrated to a depth of three meters and performed a full chemical analysis to verify the presence of ice.

International Lunar Quinquennium (ILQ)

The discovery of ice on the Moon gave impetus to a European proposal for a five-year program of scientific missions: an International Lunar Quinquennium (ILQ), to last from 1 January 2000 to 31 December 2005. The European Space Agency sent MORO (Moon ORbiting Observatory) into lunar orbit to further explore the Moon's surface, and in February 2003, LEDA (Lunar European Demonstration Approach) soft-landed on the Moon, showing that the Europeans were also capable of participating in lunar exploration. Both the Japanese and the Americans landed lunar rovers to survey possible sites for lunar bases.

During a second phase of the project, from 2005 to 2008, pilot plants were constructed for the production of food, oxygen, and energy. In 2009 Lunox (LUNar OXygen) was founded by the American Carbotek Corporation and the Japanese Shimizu Corporation. Construction of Moon City also began that year. Since then humans have permanently lived and worked on the Moon.

GOVERNMENT

Who Owns the Moon?

Throughout history, planting a flag in unclaimed territory has meant ownership, and the United States has had its flag on the Moon since July 1969.

But the former Soviet Union was the first to crash-land national insignia on the Moon. That was 10 years earlier, in September 1959. Japan and Europe also have left their pockmarks on the Moon. Japan crash-landed the Lunar Orbiter Hiten in April 1993. The European Space Agency crash-landed its first orbiter in January 2000. By that time, lunar exploration by

Japan and the United States was already well underway. So who owns the Moon?

Space Law
As the presence of humans on the Moon increases, so must the laws governing their actions there. Space law had its origins in 1958 with the formation of the U.S. National Aeronautics and Space Administration. The following year, the United Nations created a committee to study legal issues involved in exploring and developing outer space. Between 1967 and 1976 the U.N. drafted five international treaties. These state:

"All people have equal access to outer space. An astronaut or space equipment that accidentally lands in one country must be returned to the launching country. Countries are responsible for what they launch. Everything in space must be registered."

And finally: "The Moon and other celestial bodies are the heritage of and are to be shared by all humankind."

So we all own the Moon. But don't dream about buying beachfront property on the Sea of Tranquility, though. Today's Lunar Administration carefully registers all activities and assigns grounds or craters only for general use. Within these assigned boundaries, the law does allow commercial or even private use of property. Still, land development must first meet all the requirements of the Lunar Exploration Act as well as the even tougher Lunar Conservation Act.

You can travel very comfortably on the Moon by Lunar Module. Despite the large distances between the lunar bases, the trips are short.

ON THE SURFACE

SIGHTSEEING HIGHLIGHTS

No one can see everything there is to see on the Moon in one day. Your itinerary will depend for the most part on your individual interests. A word to the wise: Plan your trips well in advance, taking into account the lunar night and day. If you set off on a whim, you might find yourself in an area where the two-week-long lunar night has just begun.

If you're traveling to the Moon for the first time, it's best to arrive between New Moon and First Quarter. The terminator, the dividing line between night and day, will just be approaching the Moon Museum. You can visit the museum before the Sun rises and then take off for other destinations during the lunar day.

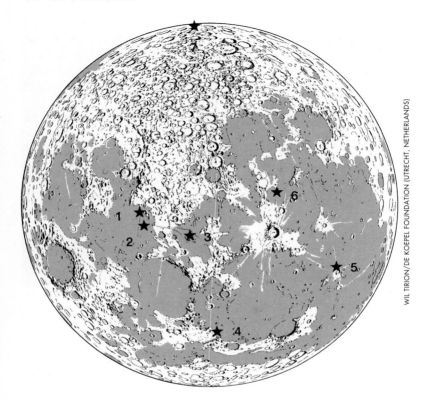

Bases on the Moon: 1 Moon Museum, 2 Lunox headquarters, 3 Lunox excavations near Mt. Schneckenberg, 4 Mont Blanc Resort, 5 Selenological Survey, 6 Moon City, 7 Water ice region in Amundsen

Begin by exploring the region around the Moon Museum. After that you might head north to the Lunox oxygen and concrete production plant. From there you could take the monorail to see Ariadaeus Rille and the excavations at Mt. Schneckenberg. In the meantime, the Sun will have risen there as well. From Mt. Schneckenberg you can travel to Mont Blanc Resort. There is so much to do in the resort itself and the area is so beautiful that you should stay at least a week.

Of all the memories visitors take back to Earth, the resort's highly sophisticated dome is one they never forget. Hikes to the edge of the dome to view the desolate lunar surface (no spacesuit required) are very popular —don't pass by this opportunity. Inside the glass is a virtual paradise with a park and splashing streams; outside is a dead and airless rocky desert that is either blazing hot under the rays of the sun or deeply frozen during the lunar night.

From Mont Blanc you will probably want to head to Moon City, with a short stop in Schröter's Valley on the way. While in Moon City you can visit the many historic landing sites in the area and the lovely Copernicus Crater. If time permits, you should definitely consider visiting the South Pole to see the famous Mountain of Eternal Light, stopping at the Straight Wall on the way.

RECREATION

SPORTS

Track and Field

Even the laziest couch potato can perform some seemingly great athletic feats on the Moon. Since its low gravity doesn't affect your normal muscular strength, you can lift six times more mass on the Moon than you can on Earth.

High-jumpers also can theoretically jump six times higher on the Moon than on the Earth. Still, getting your body over the bar is a little trickier. Before the jump, the human body's center of mass is about one meter above the ground. So athletes have to bring their bodies over the bar only from that point. To calculate the jump on the Moon, first subtract the one meter from the jump on Earth, multiply the difference by six, and then add back in the one meter to the result. An athlete who can manage a bar placed at 240 cm on Earth can jump a bar at about 940 cm on the Moon.

The shot put is even more striking. On Earth, a heavy shot weighing 7.25 kg thrown at a speed of 14.2 meters per second from a shoulder height of two meters can travel a distance of 22.47 meters. On the Moon, the same shot would land no less than 126.45 meters away.

Low gravity isn't always helpful, though. You can't run very fast on the Moon. Every step you take with the muscular strength you would use on Earth will propel you six times higher on the Moon, and it will take you six times longer to get your feet back on the ground. You might manage a few giant steps and hops, but very little else.

Because of these spectacular performances, athletic activities are permitted only at designated areas on the Moon. For sports enthusiasts, a gym in Moon City called the Hall of Diminished Gravity (affectionately known as the Dim Gym), has been specially designed to accommodate lunar athletic feats.

In addition, the Mont Blanc Resort complex has a swimming pool with 30- and 60-meter diving towers, and a lunar golf course, where the players are obliged to follow special rules. In the vicinity of the crater Plato, there's a driving range where you can practice real golf swings.

Golf

At the request of the many visitors, the owners of Mont Blanc Resort built a golf range in the middle of the Christa McAuliffe Memorial Park. But golfers beware: Lunar golf is a peculiar cross between miniature golf and the "pitch and putt" holes favored by beginning golfers. The rules are different here.

DRIVE YOUR OWN MOON ROVER

TOO EXPENSIVE, A VACATION ON THE MOON? THEN TRY VISITING IT BY TELEpresence. You can drive the remote-control rover put on the Moon in late 1997 by Lunar Corp. of Arlington, Virginia.

The lunar rover, which transmits live images from its stereo TV cameras, landed near the Apollo 11 landing site and is still performing Neil Armstrong's nostalgic "MoonTrack" tour. The Apollo landing sites are popular destinations. An early operator even used the rover to find the grave of the lost Soviet rover Lunakhod.

You can pilot the lunar rover at interactive stations at science fairs and museums for normal theme-park attraction prices. Also, check out car dealers for free trial runs. "We'll let you test drive the lunar rover if you test drive our latest model," they sometimes advertise.

To begin with, never hit the ball with a full swing. If you were to use a 12° driver as you would on Earth, the ball would land 800 meters away. In the park's artificial atmosphere, aerodynamic balls have a lot of lift. Besides that, the gravity is six times lower than on Earth, and people who fail to complete a full swing are likely to launch themselves. Alan Shepard nearly did that during the second manned mission to the Moon, Apollo 14 (1971).

Shepard, wanting to be the first golfer on the Moon, brought three golf balls that he planned to "drive" across the Moon's surface. He strapped a six iron to the bottom of a sampling instrument and told Mission Control to watch. According to NASA lore, the balls flew off into the distance and landed several kilometers away. In the lunar void, a golf ball hit at a speed of 200 km per hour at a 45° angle would travel 1.9 km. But the truth is that Shepard buried the ball on the first swing. His second swing knocked the ball a few feet before it came to a dead stop. Mission Control, providing commentary, said that the shot was a slice. His last shot was more successful, and the ball disappeared in a graceful arc in the direction of a crater. "Miles and miles and miles," cried the triumphant Shepard. He left a ball for future golfers (see "Astrotrash" in the "Environmental Issues" section in the Introduction).

To prevent falls, you may putt only with chippers. No lunar golfer's bag would be complete without an ordinary putter and a dust wedge, a club for hitting the ball out of dust traps.

The **Shepard Driving Range** lies near the landing platform in Plato Crater. Here you can drive luminescent golf balls to your heart's content. Unlike the lunar golf range in Mont Blanc Resort, you can swing as you would on Earth. To keep you from falling, an elastic cable is hooked to the back of your moonsuit. You can aim at two crater pits 1.6 and 1.8 km away. After a half-hour session, it's time to go look for the landing sites of the numbered and multicolored balls.

Swimming

You'll find the Mont Blanc swimming pool next to the lunar golf range. It is surrounded by a glass screen to keep the water, which splashes six times higher than on Earth, inside the pool area. High-divers take advantage of 30-meter and 60-meter diving platforms. A diver takes a little more than six seconds to hit the water from the 30-meter board and nearly nine seconds from the 60-meter board. Because both of these jumps last two and a half times longer than on Earth, it's a real treat to see how many gainers and twists experienced divers can accomplish.

Solar Sailing
Those who don't care for golf or swimming can try their luck at solar sailing. The lightweight "go-carts" lined up near the Plato Crater landing platform are equipped with solar-powered "sails" that can be tilted in all directions. Sailors can explore the entire sunlit part of the crater by solar energy. The go-carts are equipped with a satellite navigation system so that no one can get lost beyond the flat horizon.

The crater, with an inside diameter of 87 km, is as slippery as a sheet of ice. This makes it possible to speed from the middle of the crater to the crater wall and back again in record time. Most solar sailors head for Plato Zeta, the collapsed southwestern wall that stretches 22 km and towers 1,300 meters above the crater floor.

EVENTS

Eclipses of the Earth and Sun
If you're lucky, you'll see one of the eclipses of the Earth and Sun visible on the near side of the Moon. During "terrestrial eclipses," the Moon moves between the Earth and the Sun and casts its umbra on the Full Earth. You can see a dark spot on the Earth and watch it crawl from west to east. At least two and no more than five of these eclipses occur every year.

But the solar eclipses, in which the Earth blocks the Sun, are even more spectacular. Solar eclipses occur two to four times a year. Usually they are partial eclipses, but Moon visitors should try to catch one of the occasional total eclipses, in which the Earth covers the Sun completely.

During a total eclipse, the Earth glides in front of the Sun. The sunlight deflected by the Earth's atmosphere creates a reddish-orange rim around the Earth. As the eclipse progresses, the rim grows until it completely surrounds the Earth. At that moment, a soft, reddish-orange glow bathes the entire lunar landscape. And the night side of the Earth is watching a total lunar eclipse.

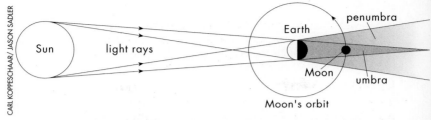

Solar Eclipse: When the Full Moon moves through the Earth's umbra, a lunar eclipse occurs on Earth. But on the Moon, the Earth blocks the Sun, so observers can see a spectacular solar eclipse.

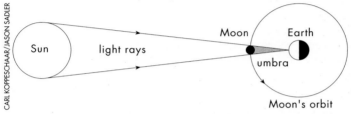

Terrestrial Eclipse: When the New Moon moves in front of the Sun, a solar eclipse occurs on Earth. However, on the Moon, watchers can observe a "terrestrial eclipse" as the Moon moves between the Sun and the Earth and casts its dark umbra on a section of the Earth.

On very rare occasions, no reddish ring surrounds the Earth, and it gets very dark on the Moon. This means that a volcano has recently erupted on the Earth, sending a lot of dust into the atmosphere. This volcanic dust floats for years in the upper layers of the Earth's atmosphere and intercepts almost all the deflected sunlight.

Since total solar eclipses on the Moon are so beautiful to behold, the chart on page 75 lists the dates of the upcoming eclipses. If it's at all possible, try to make your trip to the Moon coincide with a total eclipse of the Sun. It's a truly unforgettable experience.

But be careful when you observe an eclipse. The intensity of the light will burn your eyes. If you're watching from inside a dome, be sure to pick up a pair of EclipseBans, available at all lunar bases.

People outside on the surface should install a "solar visor" so they can look directly at the partial phases of the solar eclipse. During totality, when the Earth completely hides the Sun and only a bright orange ring surrounds the planet, you can flip up your solar visor and look through

> ## THE WOMAN IN THE MOON
>
> THE ANCIENT CHINESE HAD A LEGEND ABOUT THE WOMAN IN THE MOON. As they tell it, one of the captains of Emperor Yao's squad of archers and bodyguards was a man whose abilities bordered on the miraculous. More than once during a lunar eclipse he had rescued the Moon from the dangerous claws of the monster threatening to swallow it. As a reward for his good deed, a goddess presented him the herb of immortality. However, Hung Ngo, the archer's wife, had her eye on that magic herb. Why the archer did not choose to share it with his wife we are not told. We only know that she absconded with the precious herb. Heeding the advice of a sorcerer, Hung Ngo sought refuge on the Moon. She was no sooner there than the goddess who had given her husband the herb changed her into a toad. And this is why even today, when the rest of us mortals look at the dark spots on the Moon, we can still make out the shape of a toad.
>
> ### Mama Qilla
> To the ancient Incas, Mama Qilla, the Moon, was the supreme goddess, ruler over the feminine just as the Sun governed the masculine. Inca women worshipped her, and even in the post-Inca Andes they sought her help in childbirth and conception. But the Moon—both sister and wife of the Sun—was also to be feared. During an eclipse, she could turn a woman's tools into ferocious animals.

your helmet, as you normally would. Be sure to turn away the moment the Sun reappears and lower your solar visor again to watch the end of the eclipse.

Eclipse parties are arranged at the moonbases for people who want to watch from the outside (and who doesn't want to?). The organizers can help you rent solar visors and get them installed. You can join an eclipse party, which includes a walking tour, lecture, and visor rental, for about S$20.

Moonquakes
During the Apollo missions, the astronauts stationed seismometers on the Moon, and these continue to measure moonquakes. Most of the roughly 3,200 quakes a year begin at depths of 600-950 km. These quakes are strongly influenced by the gravitational pull of the Earth, occurring most

TOTAL SOLAR ECLIPSES ON THE MOON

Year	Date	Type*	Year	Date	Type*
2021	26 May	Total	2028	12 Jan.	Part
2021	19 Nov.	Part	2028	6 July	Part
2022	16 May	Total	2028	31 Dec.	Total
2022	8 Sept.	Total	2029	26 June	Total
2023	28 Oct.	Part	2029	20 Dec.	Total
2024	18 Nov.	Part	2030	15 June	Part
2025	14 March	Total			
2025	7 Sept.	Total			
2026	3 March	Total			
2026	28 Aug.	Part			

* Part: visible on part of the Moon
Total: visible on the entire near side of the Moon

frequently when the Moon is nearest to the Earth and the stress in the Moon's crust is at its greatest, and to a lesser extent when the Moon is farthest from the Earth. Another type of moonquake has been associated with solar heating. Weak signals of such moonquakes have been recorded about 48 hours after local sunrise and appear to result from the thermal stresses induced by the temperature changes that affect the regolith.

According to the accounts of the Spanish conquistadors, the Indians in Peru panicked during a lunar eclipse. They believed that the Moon was dying and would destroy the Earth when it fell from the sky. For the duration of the eclipse, they made loud noises and beat their dogs to make them howl, all the while shouting "Mama Qilla," or "Mother Moon."

But the strongest moonquakes are caused by meteorites. On 17 July 1972, a meteorite with a mass of 1,000 kg crashed into the far side of the Moon. Seismic recordings of this event indicate that part of the Moon's core must be molten, or at least partly molten.

ACCOMMODATIONS AND FOOD

WHERE TO STAY

Though the Japanese began designing lunar cities in the 1990s, the colonies have just recently been built and can support only a limited number of visitors. Tourism is becoming big business on the Moon, though, and more hotels are scheduled to open. Once those already under construction are completed, the accommodation crunch will ease. For now, tourists must book their hotels well in advance. The tourist offices and travel agencies on Earth will gladly help you with bookings.

Moon City, with just two hotels, has decidedly fewer accommodations than the Mont Blanc Resort. At the **Nikkei,** in the neighborhood of the Hall of Diminished Gravity, you get good value for your selenes. The hotel rooms are first class, with all the amenities, and the rates are reasonable.

The **Grand Shimizu** downtown, more luxurious but also more expensive, is typical of the early days of lunar colonization, which means that it is almost completely hidden underground. It may be the perfect experience for the person who has done it all. But during my last stay there, I found the service to be highly impersonal.

The **Mont Blanc Resort** is just that, a resort, and the several hotels are comparable in services and rates. All offer golf and solar-sailing packages, guided tours of the surface, and if the time is right, solar-eclipse excursions. Take your pick of the **TraveLunar Lodge, Alpine Hotel, Starbright,** or the **Seaside Resort.**

WHERE TO EAT

Moon City boasts a variety of restaurants: **Far Earth** is housed in a glass dome that permits an excellent view of Earth. Its four-star chefs prepare, among other specialties, steaks that are imported from Earth and served with lush, green lunar salad. **Kenji's** serves Japanese food only. You can find **American-style fast-food** restaurants at the Hall of Diminished Gravity. MacMoon is the most popular. Both Moon City hotels have restaurants featuring international cuisine.

Most of the restaurants in Mont Blanc are attached to the hotels. Being a chef at a Moon resort is so prestigious that competition for the jobs is fierce. As a result, the menus reflect the influence of the top graduates of Earth's finest cooking schools.

Moon cuisine specialties and favorite imports, available not only at restaurants but at most other shops, are detailed below.

FOOD AND DRINK

Cheese
Naturally enough, cheese is one of the most popular foods available on the Moon. Tradition has long had it that the Moon is made of green cheese, because a green (in the sense of new) cheese has the shape and color of the Moon. A popular menu item at moonbases everywhere is the Moon Rovers' Lunch, which consists of a hunk of bread, a hunk of cheese, and a pint of moonshine (see below).

Moonshine
The favorite local hooch is brewed from the milky juice of the locally cultivated **moon plant** *(Sarcostemma brevistigma),* a vine brought here from East India.

Moon Pies
From their humble beginnings in an east Tennessee bakery back in the 20th century, these chocolate-covered marshmallow graham-cracker cookies spread first across the American South, and then across the country and around the world. It was only natural that the construction workers for the

first moonbases would bring a supply. Moon Pies are available all over the Moon, though they still have to be imported because bakers can't produce marshmallow in the Moon's low gravity.

Moon Cakes
On Earth, the early association of the Moon with life and fertility (see special topic "The Influence of the Moon" in the Introduction) inspired several customs. One, referred to by the prophet Jeremiah, involved the baking of Moon cakes. "Seest thou not what they do in the cities of Judah and in the streets of Jerusalem?" he said. "The children gather wood, and the fathers kindle the fire, and the women knead their dough, to make cakes to the queen of heaven." The "queen of heaven" was Ashtoreth, a fertility goddess associated with the Moon. Bakers on the Greek mainland and in Egypt, India, and China made similar cakes. In the 19th century cakes honored the Moon in both England and China. Scholars have suggested that the "hot cross buns" of Christians' Easter are an old pagan custom connected with Moon worship.

The Chinese invented the Moon cake to honor the Moon during the Autumn Moon Festival. They are very popular on the Moon, and you should be able to find them on all the moonbases. But in case you can't, the recipe follows. You should make this old Earth recipe with Chinese Moon cake molds, which imprint the traditional chrysanthemum pattern and Chinese characters on them. But you can easily mold them yourself: they should be about seven to eight cm in diameter. This recipe makes 16 cakes.

> **Pastry:** four cups flour, four tbsp. brown sugar, half tsp. salt, one-half cup margarine
>
> **Filling:** two tbsp. each of peanuts, sesame seeds, and walnuts or pine nuts; two tbsp. boiled chestnuts or blanched almonds; two tbsp. chopped sultanas or other dried fruit; two tbsp. apricots; and four tbsp. brown sugar, two tbsp. margarine, two tbsp. rice flour
>
> **Glaze:** one egg, one tsp. sesame oil

Preheat oven to 400° F or 200° C. Sift flour, sugar, and salt together. Chop the margarine into pieces and cut into the flour until crumbly. Add enough hot water (about one-half cup) to make a dough. Cover with a cloth. Roast the peanuts in a hot pan for two minutes. Add the sesame seeds; put a lid on or they'll jump out of the pan. Roast for two more minutes. Put the peanuts and seeds in a processor or blender and grind with other nuts. Add to the rest of the filling ingredients and mix. Roll out the pastry on a floured board. Cut rounds to fill the mold or make little pie shells. Rub the mold with margarine and spread pastry over the bottom and sides. Put in a tablespoon of filling. Press down gently. Wet the edges of the pastry and cover with another round to make a lid. Seal, and remove from the mold. Put all the cakes on a greased baking sheet. Beat the eggs and sesame oil together and brush each cake with this mixture. Bake about 30 minutes until they're golden brown.

TRANSPORTATION

GETTING THERE

In his 1865 book, *From the Earth to the Moon,* Jules Verne described a trip to the Moon with uncanny foresight. The main character traveled to the Moon in a kind of hollow torpedo, which circled the Moon and returned to Earth. In Verne's day, space travel wasn't even a remote possibility, yet he had the space capsule landing in the water. It wasn't until a century later that the Apollo missions splashed down in the Pacific Ocean, making the kind of "wet" landing described by Verne.

These days the trip is a lot more comfortable. First, the Single-Stage-To-Orbit (SSTO) vehicle brings you to Hilton Space, the satellite hotel in a low-altitude Earth orbit from which flights to the Sea of Tranquility depart. Here you'll board the Lunar Transfer Vehicle for the longest leg of your journey. Soon after launch, you'll see the hotel disappear from view, becoming a mere pinprick of light in the distance. In less than minutes, you can no longer distinguish it from the blue and white shining Earth. Through the observation window of the Lunar Transfer Vehicle, you can see the silhouette of the Moon getting larger and larger.

You'll dock at the space station orbiting the Moon, and from there a landing craft will take you to the lunar surface. Provided your connecting flights go smoothly, you can be on the Moon in two and a half days.

Jules Verne wrote his book From the Earth to the Moon *in the 19th century. The spaceship depicted here will take three astronauts to the Moon.*

  (opposite) Remote-controlled robots called "teleprospectors" are used to find minerals on the Moon. (Pat Rawlings/NASA)

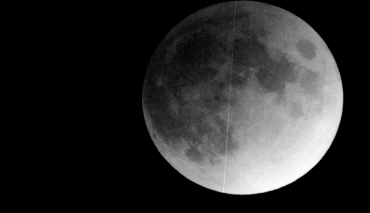

(above) A total lunar eclipse is one of the most beautiful phenomena that can be observed from the Earth. During a total eclipse, the Moon is illuminated by a faint, reddish-orange glow. (Carl Koppeschaar) (below) Long ago, the Moon was a lot closer to the Earth. They are moving 3.5 cm a year away from each other. (Don Dixon)

What the Moon looks like from outer space, according to the Tintin comic book Explorers on the Moon. *The caption reads "Is that really the Moon? . . . That funny ball riddled with little holes?"*

It's 2020, and 23 years since the space agencies of Earth launched the joint Return to the Moon project. Their first step was to send robotic probes to check out prospective sites for lunar bases. From the beginning, one thing was sure: the first base would be built in the Sea of Tranquility, since this is where, in 1969, human beings first set foot on the Moon. Besides, the Sea of Tranquility is on the near side of the Moon, allowing the base to maintain constant radio contact with the Earth. A base could be built on the far side of the Moon only later, after a network of communication satellites had been set up.

The Approach to the Moon

During your trip, the image of the Moon changes. Up to a distance of 10,000 km from the Moon, you can still see virtually all of the surface normally visible from the Earth. Only a small piece along the edge is invisible, but you probably wouldn't notice that unless you took the time to compare what you see to a detailed map of the Moon. But after that the image changes drastically. At a distance of 5,000 km, you can see 97% of the lunar surface facing you. At 1,000 km only 77% is still visible, at 500 km only 63%, and at 200 km only 44%.

Unfortunately, the Lunar Transfer Vehicle makes its approach on the night side of the Moon. Still, the Moon isn't enveloped in total darkness; there's a kind of grayish glow—the sunlight reflected from the Earth. The Earth reflects so much sunlight in outer space that it also casts a faint light on the surface of the Moon. Under a "Full Earth," as seen from the Moon, the earthlight is extra bright.

As the Lunar Transfer Vehicle soars past the western edge of the Moon, the Earth sets and the night begins. It's kind of scary, seeing a gaping "nothingness" on one side and a brilliant star-studded sky on the other. In the meantime, you'll be cruising at an altitude of 200 km. The visibility increases until you can see 797 km in all directions, which means that the Sun will rise in two minutes.

Sunrise, at 200 km above the surface of the Moon, is an event like nothing you've ever seen on Earth. The Moon has no dawn or dusk, because it has no atmosphere to deflect and diffuse sunlight. Suddenly you see dazzling light and a thin jagged line along the edge of the Moon. Tall peaks and mountain ranges cast inky black shadows. It takes some time before you can make out entire craters.

It comes as a shock to see the lunar surface loom up at such close range. At first sight, it looks like a battlefield that has just been bombed.

On the far side of the Moon, the Lunar Transfer Vehicle docks with the space station, which is permanently orbiting the Moon. This lunar space

VISIBILITY ON THE MOON

DISTANCE	VISIBLE HORIZON (KM)	VISIBLE SURFACE AREA (%)
384,000 KM	2722.2	50.00
100,000 KM	2700.4	49.99
10,000 KM	2471.8	49.45
1,000 KM	1534.8	38.64
100 KM	575.9	16.27
10 KM	186.0	5.34
1 KM	58.9	1.70
100 M	18.6	0.54
10 M	5.9	0.17
1 M	1.9	0.05

A landing craft makes its descent to the Moon.

station takes about two hours and seven minutes to complete one orbit. After passing through an air lock, you'll wind up in the transit area, where you'll pick up your connecting flight to the Moon below.

Arriving on the Moon

For passengers bound for the Moon Museum and TraveLunar Hotel, the time has come to switch to one of the smaller landing crafts. These Lunar Landers can accommodate two crew members and four passengers. Once the space station passes the beautiful Langrenus walled plain, the Lunar Lander disengages. It gradually comes under the influence of gravity, so that the last part of the journey is mostly in free fall. You land on the Moon in the southwestern corner of the Sea of Tranquility, just a few kilometers from the spot where the astronauts first set foot on the Moon.

TRANSPORTATION ON THE MOON

LAUNCH VELOCITY		TRAVEL TIME	DISTANCE (KM)	ALTITUDE (KM)
M/S	KM/H	(MIN:SEC)		
100	360	1:27	6	1.5
250	900	3:38	39	9.6
500	1800	7:15	155	38.6
750	2700	10:53	349	86.8
1000	3600	14:30	620	154.3
1250	4500	18:08	969	241.1
1500	5400	21:45	1395	347.2

While the retro-rockets raise a cloud of dust, the Lunar Lander comes to a halt on its shock-absorber legs. A moon buggy (officially known as a Lunar Roving Vehicle) approaches in another cloud of dust, and you can hear the voice of the captain in your space helmets: "There's the welcome committee, folks. Enjoy your visit."

While the Lunar Lander is being readied for the return flight, your moon buggy drives off in the direction of the Moon Museum. The museum, in the glow of floodlights, consists of a gigantic pressurized dome, like the ones used on Earth for traveling exhibits. A walkway leads to the building behind it, which is partially underground. This is the lunar base, which also serves as a hotel for visiting tourists.

GETTING AROUND

Thanks to the low gravity, transportation on the Moon is simpler and more energy-efficient than on Earth. For relatively short distances, travelers can avail themselves of moon buggies and the monorail. Lunar Modules provide transportation between the various bases. You can also transfer to another mode of transportation at the space station orbiting the Moon.

The LMs are equipped with reaction-control thrusters, so that the pilot can keep the module on a horizontal course. Short bursts of retro-fire make mid-course corrections. The LMs are launched and land at a 45° angle, a trajectory that optimizes the distance and is therefore the most energy-efficient. The launch requires only a short combustion of the main rockets. Since the Moon has no atmosphere and therefore no friction, your flight will continue without interruption in a parabolic trajectory. All the pilot has to do to land is to maneuver the module into an upright position and slow its speed of descent by firing short bursts of retrofire.

For longer trips, such as those between the far and near sides of the Moon, you'll fly in a lower trajectory with a higher starting velocity. Or you can even orbit the entire Moon. The orbital velocity for circular motion directly above the lunar surface amounts to a mere 1.68 km per second (6,048 km per hour). Since the Moon has a circumference of 10,920 km, you can circle the entire Moon in 1 hour and 48 minutes.

The chart lists the time it takes the Lunar Modules to travel the given distances at various launch velocities and altitudes.

You can arrange individual transportation by chartering Lunar Modules from the two carriers operating on the Moon: Translunar Flights and Moonlink. All flights are short, so you can go anywhere you like on the Moon. Prices vary according to destination.

INFORMATION AND SERVICES

OFFICIALDOM

The Lunarians welcome visitors from all countries. Since the Moon is considered the common heritage of humankind, visas are not required. Only a limited number of travelers may visit at one time, so be sure to plan accordingly.

HEALTH AND SAFETY

Inoculations
The Moon is free of viruses and disease. Inside the domes and city the air is kept at a constant comfortable temperature so it's next to impossible even to catch a chill. To make sure the Moon stays free of germs, your inoculations must be up to date before you visit.

While scientists knew that the Moon was covered by a layer of dust, they didn't know how deep it was. Before the first mission they feared that the landing craft would be swallowed up in a deep layer of dust. However, the dust proved to be no more than a few centimeters thick. It clings together so much that shoes leave clear footprints.

Radiation

The Earth's atmosphere protects Earthlings from the Sun's deadly ultraviolet rays and slows down meteorites that would otherwise collide with the planet at a high speed. The Earth's magnetic field also serves as a shield against the "solar wind": a constant flow of charged particles from the Sun. Lacking the protection of an atmosphere, all of the Moon's permanently manned bases are covered by a layer of lunar concrete two meters thick to shield them from solar wind and other forms of cosmic violence. Moon buggies, Lunar Modules (LMs), and other transportation vehicles have been fitted with a nine-cm-thick layer of aluminum to keep the amount of radiation to a minimum.

Moondust

Constantly exposed to extreme cold and extreme heat, the upper layer of the Moon's surface cracks and eventually crumbles into dust. The dust blanket acts to insulate the subsoil, which remains firm. The dust is never deeper than a couple of centimeters, so it's possible to walk around everywhere on the Moon.

The lunar dust creates a very sticky problem, though: the solar wind charges the dust particles with static electricity, so that the moondust clings to everything, including your moonboots and the folds of your moonsuit. You can easily get rid of the dust by removing its static charge; all you have to do is to step on a noncharged surface. Electrostatic discharge devices are stationed at the entrances and exits to every lunar base and in lunar vehicles operating on a closed life-support system. Once it loses its static charge, the dust either falls to the ground or can be suctioned off.

MONEY

Major credit cards and traveler's checks are accepted on the Moon. But it may be convenient to have some change on hand as you stroll through Mont Blanc Resort or down the underground streets of Moon City. The local currency is the **selene,** named after the Greek goddess of the Moon.

The selene is divided into 100 cents. The 5-, 10-, and 25-cent coins are made of nickel. Bronze coins come in denominations of 1, 5, 10, and 25 selenes, and small titanium coins come in denominations of 100, 250, and 1,000 selenes. Paper money is not used on the Moon because at its reduced surface gravity, even the tiniest draft would easily blow the bills away.

SPIN A COIN

LUNARIANS FROWN AT THE EARTH PRACTICE OF FLIPPING COINS TO SETTLE BETS or make a choice, not because of superstition but because of the possibility of cheating. The flipped coin falls so slowly that the person who catches it may be able to see whether it's going to be heads or tails. If you should need to toss, place a coin on a table or any other flat surface and spin it on its edge. Because of lower surface gravity, the coin will fall only slowly, and watchers can't easily predict on which side it will land.

Tourists from Earth have also nicknamed the selene the "yollar," as it is coupled to both the Japanese yen and the U.S. dollar. The exchange rate is 1 selene (S1) = ¥100 = US$1 (June 2020).

COMMUNICATIONS

Because the Moon lacks an atmosphere, the lunar world is eerie and silent. No wind, no snow, no rain, no mist, no clouds: It has none of Earth's everyday changes. And sound doesn't travel, which means that communication can take place only by means of radio contacts. Space helmets are equipped with built-in transmitters and receivers, which are linked to a network of communication satellites. Direct radio contact is possible only at a short range on a flat surface, within a radius of about 2.5 km.

The Moon boasts two local publications: Moon City's daily *Lunarian Tribune,* and the Mont Blanc *Herald-Gazette,* a weekly. Both are available at shops around the Moon, as are the world's major newspapers, which are beamed by satellite from Earth.

TOURIST INFORMATION

The sightseeing exhibit near the entrance of the Moon Museum doubles as a **tourist information office.** Here you can find all the information, including hotel brochures, timetables, and maps, that you need to help you enjoy your vacation, and the staff can help you plan your trips and book your flights.

WHAT TO TAKE

Moonsuits

Modern moonsuits protect you against the extremes of heat and cold for six hours. After that, it's time to recharge the batteries and refill the oxygen tanks in the Portable Life Support System (PLSS). The PLSS regulates the temperature in the suits by heating or cooling the liquid that circulates in tubes in one of the underlayers of the garment. Fresh PLSSs are always available at the entrances and exits to the lunar bases. In case of an emergency, additional PLSSs are stored on every transportation vehicle.

The moonsuits, with their 21 protective layers, likewise will shield you from an average dose of radiation and from falling meteorites. The suits are also tear-proof, so don't worry about running into sharp or falling rocks. Besides that, it's practically impossible for someone in an inflated moonsuit to break an arm or a leg. As for the helmet, it's so strong you couldn't dent it with a sledgehammer.

Clothing

Though the PLSS completely regulates the temperature inside the space suit, you wear special lightweight clothing underneath. You can buy these "undergarments" at shops on the Moon. (You can also buy them on Earth, but prices are the same in both places.) People continue to walk around in these clothes after getting out of their moonsuits in the regulated environment of Moon City and other settlements. Pay attention and you'll spot various fashion trends.

People who aren't going to wear their moonsuits for a long time usually also bring light summer clothes. Americans favor Bermudas and Hawaiian shirts; many other visitors are more formal and usually dress up for dinner in their hotels.

WEIGHTS AND MEASURES

Time

The Moon, which revolves around the Earth, provides Earth-dwellers a unit of time that is longer than a day. One entire revolution of the Moon is a "month," a word derived from the Middle English word for moon, *moneth*.

The same side of the Moon always faces the Earth. Consequently, if you're on the near side of the Moon, you will always find the Earth at the same place in the sky. The Earth is like a constantly visible clock, whose phases can be used to represent a lunar day, which lasts about 29.5 Earth days: Full Earth means that it's midnight at the middle of the side facing the Earth; morning begins at Waning Earth; New Earth means it's noon; and the day winds to an end through Waxing Earth. *(continues on page 92)*

LUNAR CALENDAR

IF YOU SCAN THE WESTERN HORIZON ON A CLEAR NIGHT A DAY OR TWO AFTER New Moon, you will see a gossamer-thin sliver, the new Crescent Moon. Thousands of years ago, the new month began when observers sighted this crescent. Excavations have brought to light Babylonian clay tablets recording these narrow Crescent Moons as long ago as 750 B.C. But there is also evidence that the Sumerians were tracking this phenomenon a thousand years earlier in Mesopotamia.

Month

The Babylonians knew that the average span from one New Moon to another was 29.5 days. The clay tablets indicate that at the end of every 29th day, observers were posted to watch for the new Crescent Moon. When it was sighted, the new month began immediately. When the crescent could not be seen (because of bad weather or because the actual period of time between two successive New Moons can vary between 29.25 and 29.75 days), the new month began the following evening. Therefore, the months lasted either 29 or 30 days.

Thousands of years ago in Mesopotamia, observers were on the lookout for the Crescent Moon that signaled the beginning of a new month. The clay tablet on the right shows the calculations of the New Moon and the dates of the Crescent Moon in cuneiform.

Ramadan

Thousands of years later, the custom of beginning the new month with the sight of the narrow crescent Moon has far from died out. Both the Jewish and the Islamic calendars are still closely related to the Babylonian calendar. Certain holidays can begin only when observers have actually sighted the Crescent Moon.

This practice may occasionally lead to confusion. For example, Ramadan, Islam's most holy month, might conceivably begin on different days in different places. An Islamic community relies on the testimony of at least two independent witnesses who have seen the new Crescent Moon. So if it is partly cloudy, three different cities in relative proximity might claim three different dates: today, yesterday, and tomorrow.

One disadvantage of the Islamic lunar calendar is that it never fits exactly in a solar year. Twelve lunar months last an average of 354 days. This means that the Islamic months and holidays gradually move through all four seasons. The Koran expressly forbids the introduction of leap years. The Jewish calendar doesn't have this problem. Its normal year is made up of 12 lunar months, and from time to time a leap year is added, so that there are 13 lunar months in that year.

Julius Caesar

Since Julius Caesar reformed the calendar in 44 B.C., the Western world no longer uses a lunar calendar. Annoyed by the disorder resulting from the insertion of extra months, Caesar introduced a calendar based on the solar year—on the course of the Sun and therefore on the seasons. In the Middle Ages, Pope Gregory XIII slightly modified this Julian calendar. But the principle of using leap years, giving February 29 days every four years, has remained the same.

Easter

The introduction of the solar year did not mean that the Moon's movements were no longer significant. Many Christian holidays are still celebrated according to the phase of the Moon. Easter, for example, falls on the first Sunday after the Full Moon that occurs on or next after 21 March (the ecclesiastical beginning of Spring). Therefore, the earliest Easter can fall is on 22 March (provided the Full Moon occurs on Saturday, 21 March). The latest it can occur is on 25 April (if the Full Moon falls on 20 March, the

(continues on next page)

next Full Moon will be on Sunday, 18 April). Because the date of Easter jumps back and forth, the other holidays reckoned on the basis of Easter, such as Palm Sunday, Ash Wednesday, Ascension, and Pentecost, will also fall on different days of the year.

Harvest Moon
Everyone who lives in the countryside is familiar with the so-called harvest Moon. Because the Moon revolves around the Earth, it usually comes up an average of 50 minutes later each day. But in September the Moon passes the vernal equinox (the point at which the Sun crosses the celestial equator in springtime).

Like the Sun, the Moon then climbs higher in the sky and therefore appears relatively earlier on the firmament. For several days in a row the Moon comes up at about the same time. In the Northern Hemisphere, this is known as the harvest Moon because the farmers have more light in which to harvest their crops. No harvest Moon appears at the equator. Southern latitudes experience the phenomenon in springtime when it passes the autumnal equinox (the point at which the Sun crosses the celestial equator in springtime).

You'll also refer to the three clocks in the corridor of your hotel. The first clock gives the **phases of the Moon,** as seen from Earth. Say that at the moment it reads 0.25. Since 0 denotes the New Moon, 0.5 the First Quarter and 1.0 the Full Moon, it's not hard to figure out the current phase—it's now halfway between New Moon and First Quarter.

The second clock displays the **local time** on the Moon. It is now N-00.11.56.02. This is totally incomprehensible at first, but it stands for 0 Earth days, 11 hours, 56 minutes, and two seconds before the local sunrise. Since the period between two consecutive New Moons averages 29.5306 Earth days, one lunar day—from sunrise to sunset—is therefore 14 days, 18 hours, 22 minutes, and two seconds long. So counting from sunrise to sunset, the clock runs from D+00.00.00.00 to D+14.18.22.02. During the equally long lunar night, the clock runs backward from N-14.18.22.02 to N-00.00.00.01.

The third clock indicates the date and time in **Greenwich mean time (GMT)** on Earth. Fortunately, this one doesn't take any getting used to. It simply reads: 25 June, 6:51 p.m. GMT, in the year 2020.

Gravity

Everyone who goes to the Moon for the first time has to get used to the low gravity. On the lunar surface, gravity is only a sixth of that on Earth. Therefore, an adult weighing 78 kg on Earth will tip the scales at a mere 13 kg on the Moon. The moonsuit and the PLSS, which add another 126 kg to your weight on Earth, will help compensate. But at a total of 34 kg, you'll still feel as light as a feather.

If you want to see how lunar gravity works, you simply have to measure the time it takes an object to fall. Everything on the Moon falls a lot more slowly than on Earth. The chart shows how long it takes objects to fall on the Earth and the Moon and the speed of these objects on their way down.

There's no reason to suffer from acrophobia on the Moon. Anyone who can manage to jump from a height of two meters on the Earth will find that on the Moon 12 meters is a breeze.

HOW GRAVITY WORKS

HEIGHT (M)	EARTH			MOON		
	TIME OF FALL (S)	SPEED OF DESCENT (M/S)	SPEED OF DESCENT (KM/H)	TIME OF FALL (S)	SPEED OF DESCENT (M/S)	SPEED OF DESCENT (KM/H)
0.5	0.32	3.13	11.3	0.78	1.27	4.6
1	0.45	4.43	15.9	1.11	1.80	6.5
2	0.64	6.26	22.5	1.57	2.54	9.2
3	0.78	7.67	27.6	1.92	3.12	11.2
4	0.90	8.86	31.9	2.22	3.60	13.0
5	1.01	9.90	35.6	2.48	4.03	14.5
10	1.43	14.01	50.4	3.51	5.69	20.5
20	2.02	19.81	71.3	4.97	8.05	29.0
30	2.47	24.26	87.3	6.09	9.86	35.5
40	2.85	28.01	100.8	7.02	11.39	41.0
50	3.19	31.32	112.7	7.86	12.78	45.8
100	4.51	44.28	159.4	11.11	18.00	64.8

Many historic spacecraft lie in the vicinity of Moon City. Surveyor 3 (1967) is shown here, with the landing gear of Apollo 12 (1969) in the background.

OUT AND ABOUT

MOON MUSEUM

The Moon Museum, built in 2015 and originally known as the Apollo Museum, lies near the landing gear of the Eagle, the first manned Lunar Lander, which the Apollo 11 astronauts left behind on the Moon. (See "Astrotrash" under "Environmental Issues" in the Introduction.) Neil Armstrong's first footsteps have also been preserved in the Moon's dusty surface. To ensure that nothing will be disturbed, visitors can view these objects and instruments only from a walkway five meters above the site. To save it for posterity, the probable "first footstep" has been covered by a sheet of plexiglass.

OUT AND ABOUT   95

The museum itself is divided into five areas. In the reading room, you can find practical information about your stay on the Moon. It features displays on visibility, gravity, time, radiation, rules and regulations, how moonsuits work, and how to rid yourself of moondust, among other things that every tourist should know before venturing out to explore the lunar surface.

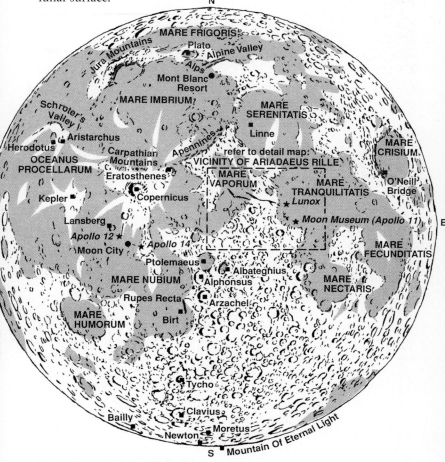

Near side sights: The highlighted features of the Moon. See detail map of Ariadaeus Rille and surroundings on page 96.

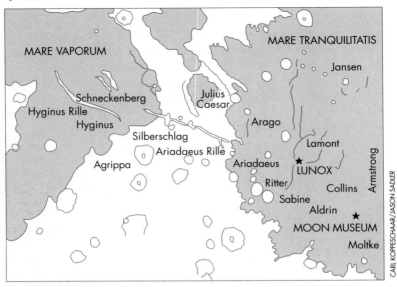

Vicinity of Ariadaeus Rille: location of the Moon Museum, Lunox headquarters and the excavations near Mt. Schneckenberg.

The room dedicated to Life on the Moon traces humankind's belief that the Moon was inhabited. Learn about Lucian of Samosata, who wrote in the 2nd century A.D. that the Moon was the final home for human souls, and the great Moon Hoax of the 1830s, in which a New York reporter had the whole world believing that the Moon was inhabited by bat people and unicorns. Did you know that the first crew to land on the Moon had to suffer quarantine for three weeks because scientists feared Moon bacteria?

The exhibits that document Space Exploration start their story in the year 3000 B.C., when a Chinese mandarin tried to reach the Moon in a spaceship fueled with 47 firecrackers. Most of the displays, though, are dedicated to the space race. Anyone interested in how the Americans beat the Russians in the frantic rush to the Moon should definitely visit this room.

The Selenography room charts the lunar surface with maps and other visual displays. It is packed with information about and descriptions of the Moon's seas, mountains, craters, valleys, and rilles. It also provides a history of the scientific debate about the origin of the Moon. Early astronomers thought the Moon was a blob that broke away from the Earth, while current theory holds that the Moon is the result of a collision—the matter isn't settled yet.

The Sightseeing room gives an overview of the most interesting spots on the Moon, both for dramatic impact and history. Besides answering any questions you might have about the most noteworthy destinations on Moon, the staff here can help you plan your trips and book your flights.

The museum is always open, though it's run by a skeleton staff until shortly before the lunar dawn. The few hours just before the terminator reaches the museum is the best time to visit, because the museum usually is not crowded then. You can study all the exhibits and get familiar with your new destinations before taking off for sightseeing trips in the daylight. Admission is free.

ATTRACTIONS

The Moon has its fair share of surface formations that stand out because of their unusual structure or beauty. Like the Grand Canyon in Arizona or some of the majestic volcanoes on Earth, they attract hordes of visitors every year.

Most of these sightseeing spots have been designated as ecologically protected areas, along the lines of a national park system. In these areas, you are absolutely forbidden to leave the roads and paths. This injunction may seem odd at first to Earth-dwellers, but there is no such thing as erosion on the Moon. Footsteps and imprints that quickly fade on the Earth because of the effects of wind and weather will remain visible in the moondust for millions of years and forever mar these places of natural beauty.

This section lists the Moon's most famous scenic spots. The "Near Side" and "Far Side" sections detail the Moon's usual tourist attractions.

THE STRAIGHT WALL

Rupes Recta, the Straight Wall, stretches out in Mare Nubium, across from Birt Crater. A sharp fault in the Moon's crust, the 96-km-long wall drops 240 meters from top to bottom. Observers from Earth used to

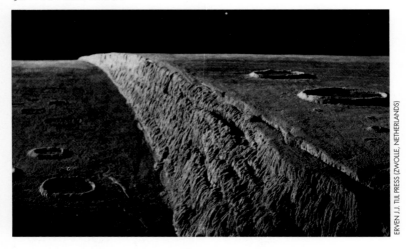

part of the Straight Wall, seen from 500 meters above the Moon

think it was a nearly perpendicular cliff between the low-lying area to the east and the highlands to the west, but its slope is never more than 40°. In any case, the Straight Wall is one of the most breathtaking formations on the Moon. You can stop to see it on your way from Moon City to the South Pole.

ALPHONSUS CRATER

The crater Alphonsus, named after the 13th-century Spanish king Alfonso X of Castile ("Alfonso The Wise"), lies northeast of the Straight Wall (14° S, 3° W). It is 110 km in diameter and 2,750 meters deep. On 24 March 1965, Ranger 9 crashed into the floor of this crater, between the central mountain peak and the northeastern wall, after relaying a series of photographs back to Earth.

On 26 October 1956, American astronomer Dinsmore Alter noticed what looked like a cloud cover above the crater's central peak. In all probability, it was a cloud of gas diffusing reflected sunlight.

Two years later, on 3 November 1958, Russian astronomer Nikolay A. Kozyrev observed this odd phenomenon again; this time the cloud had a rosy glow. Kozyrev, using the 1.25-meter telescope of the Kharkov

Observatory in the Crimean, also took several spectral photographs. He concluded that the pictures showed very weak volcanic activity.

Observers at the Selenological Survey moon station near Aristarchus saw similar phenomena on 15 June 2014 and 22 January 2015. The Moon was then in its perigee (the point in its orbit when it is nearest the Earth). The emissions were presumably related to the stress in the Moon's crust caused by the Earth's gravitational pull, which allowed gases to escape from the Moon's core. In 2016, scientists placed measurement instruments around the caldera-like summit crater in the central peak, but Alphonsus hasn't shown a single sign of activity since.

SUN AND MOON

IN MANY MYTHS THE SUN AND THE MOON ARE A MAN AND WIFE OR BROTHER and sister team who are constantly chasing each other. The Moon god or goddess was also often thought of as a sage or sorcerer. The Guarani people of South America believed that the Sun and Moon were brothers who became fish. When the Moon swallowed an evil spirit's fishing hook, he was eaten by an ogre. So the Sun collected the fish bones and revived his poor brother. His death and revival explain the waxing and waning of the Moon.

An Amazonian story tells how the Sun and Moon were once betrothed. But it was dangerous for them to marry: the Sun loved the Moon so much that his passion would set fire to the Earth, and the Moon's tears of joy would flood it. And so they decided that it would be better for them to live apart. But they had to remain at a fixed distance from each other: if they got too close they would destroy the world, and if they were too far apart they would upset the natural order of day and night.

Wei and Kapei

According to the Arecuna people of North America, the friends Sun and Moon, Wei and Kapei, were once inseparable. Kapei had a clear and beautiful face. But then he fell in love with one of the Sun's daughters. The Sun ordered her to smear Kapei's face with menstrual blood, and Wei and Kapei have been enemies ever since; the Moon's face is forever stained and he hides from the Sun.

ARISTARCHUS CRATER AND SCHRÖTER'S VALLEY

The crater Aristarchus, named after the Greek astronomer Aristarchus of Samos (circa 310-230 B.C.), is the Moon's "number one light." In the northwestern part of the Ocean of Storms (23° N, 48° W), it's the brightest object on the lunar surface.

Aristarchus is 43 km in diameter and 3,150 meters deep. To its north and northeast lies an extensive network of rilles. One of the "hot spots" on the Moon, Aristarchus stays warmer than the surrounding area during a solar eclipse (in which the Moon winds up in the dark shadow of the Earth).

In 1969 the three crew members of Apollo 11 saw a luminous glow in this crater, at the same moment that two amateur astronomers in Germany registered a flash of light. As with Alphonsus, it's apparently not an active volcano, but a fluorescence from gas clouds that escape through the Moon's crust when it's exposed to a lot of stress.

Aristarchus is also famous because it is blanketed by dark aprons of ejected volcanic debris: nine dark bands radiate from a point inside the wall. They were first observed in 1863 and have been getting progressively darker since then. Because of the changes taking place within and around this crater, the entire area is closed to all but the Selenological Survey, which is stationed here. Only nearby Schröter's Valley is open to the public.

Schröter's Valley

The famous Vallis Schröteri begins north of the nearby Herodotus Crater in an irregular depression known as Cobra Head. The valley is 10 km wide at its widest and a mere 500 meters at its narrowest. Its average depth is 500 meters. A thin rille runs down the entire length of the valley. Schröter's Valley was formed when an old lava tube that carried lava from Cobra Head collapsed.

The valley was named for Johann Schröter (1745-1816), a German lawyer and astronomer. His observations of the Moon from 1785 to 1791 culminated in the book *Selenotopographische Fragmente ("Selenotopographic Fragments")*.

LINNÉ CRATER

The crater Linné, named after the Swedish doctor and botanist Karl von Linné (1707-78; better known by his Latin name of Linnaeus), has inspired more controversy than any lunar object in the history of selenography. It's a small, bright crater with a diameter of 2.5 km and a depth of 600 meters in the western part of Mare Serenitatis (28° N, 12° E). Until the mid-18th century, Linné appeared on lunar maps as a very deep crater that was visible under all types of solar illumination.

On 16 October 1866, German selenographer Julius Schmidt (1825-84) sounded the alarm. Linné had changed. It no longer resembled a crater, but a bright white spot. It looked as if the crater had turned into a cloud. Later observers noted that the white spot displayed a shallow depression. This was assumed to be a cone, similar to a volcanic ash cone. Supporters of the volcanic theory suggested that the crater was filled with a highly fluid lava, which was flowing over the gently sloping flanks and making it impossible to see a shadow. This would explain why Schmidt saw the crater as a white spot.

American astronomer William Henry Pickering also saw the white spot, but believed it to be snow or frost. Observations made before and after a lunar eclipse served to reinforce his view. Right after the eclipse, the white spot appeared to be more than 10 km in diameter, while half an hour later, after the sunlight reappeared, it shrank again.

They all were wrong. Linné has never undergone a single change. In the 1970s the crater was photographed as part of the Apollo program. These photographs showed it to be an ordinary, relatively recent impact crater, surrounded by a bright blanket of matter tossed in the air during impact. In all probability the previous controversy was based on faulty observation.

THE O'NEILL BRIDGE

On 29 July 1953, John J. O'Neill, the science editor of the *New York Herald Tribune,* using his own telescope, discovered an arched shadow that linked Lavinium Cape with Olivium Cape in Mare Crisium (15° N, 48° W)

For a long time, O'Neill Bridge was one of the Moon's great mysteries. The bridge finally turned out to be nothing more than a shadow.

when the Sun went down. According to O'Neill the shadow was cast by a "gigantic natural bridge" 30 km long and two km wide. The press gave wide coverage to this report, especially after a UFO enthusiast claimed that the spectral photographs taken by Mount Wilson Observatory proved that the bridge was made of iron.

Despite the uproar, the bridge was nothing more than an optical illusion. Photographs taken by the Lunar Orbiters showed an ordinary gap between the two capes. At sunset the western walls of the two nearby craters cast an arched shadow in the direction of the capes. At sunrise there was absolutely nothing to be seen at that spot.

THE MOUNTAIN OF ETERNAL LIGHT

The Moon's highest mountain sits near the South Pole, where it juts out like a jaunty cap. Epsilon Peak, rising from what used to be called the Leibnitz Mountains, is actually a very tall peak in the wall of the Amundsen walled plain. Camille Flammarion (1842-1925), the French popularizer of astronomy, dubbed it the Mountain of Eternal Light because part of it is nearly always in sunlight, no matter which side of the Moon is facing the Sun.

The peak reaches 9,050 meters at its highest point, which makes it taller than Tibet's Mt. Everest. Some of the 19th-century astronomers saw it glow with all the colors of the rainbow and assumed it was covered with snow.

The foot of Epsilon Peak is always veiled in darkness, while the top is always bathed in sunlight. It looks like a spotlight hovering above the lunar surface.

The dark parts of Amundsen are the coldest on the Moon. But a great deal of activity goes on here, for this is where American and Japanese robot prospectors discovered a nearly inexhaustible supply of ice in 1997. Two months after the Lunar Prospector of NASA and the Space Studies

A nine-ton aluminum "lunar tank," designed by Boeing Corporation. The tank, equipped with robot arms and drilling equipment, allows Moon residents to go about their work protected from micrometeorite impacts and sudden sun storms.

Institute in Princeton came across the ice, the Japanese sent their Lunar A to the Moon to investigate further. It plunged a probe equipped with scientific instruments into the site, and it relayed a complete chemical analysis back to Earth. The analysis confirmed the presence of ice.

As water can be broken down into oxygen and hydrogen, the discovery of ice meant that these two gases no longer had to be transported from the Earth to the Moon. Of course it was long known that oxygen could easily be extracted from lunar rocks. But lunar rocks contain little or no hydrogen, and a mixture of oxygen and hydrogen is needed to produce a high-grade rocket fuel.

The discovery of ice provided the impetus for the joint Return to the Moon project. Several years later, in the year 2000, Earth's space agencies launched the International Lunar Quinquennium, a five-year program proposed by the Europeans. The program tested key technologies and sent a host of scientific missions to the Moon. The Moon was settled beginning in the year 2009, and Amundsen's ice is distributed to water-processing plants in the lunar bases and to the factories in Moon City that produce rocket fuel.

Long ago, a comet consisting predominantly of ice must have crashed here. In the eternally dark Amundsen Valley, the ice never felt the heat of the Sun, but awaited the arrival of humankind.

THE NEAR SIDE

Moltke Crater

The crater Moltke lies 45 km south of the Moon Museum. Its ringwall rises 300 meters above the surrounding lunar surface, and it is seven km in diameter and 1,300 meters deep—so deep that the view is hardly diminished by the curve of the lunar surface. From the middle of the crater, you can see the ringwall rise 20° above the horizon.

You can reach Moltke by moon buggy in 45 minutes. You will have to descend to the crater on foot, since the sharp crater rim will not accommodate a vehicle. Wait until 18 hours after sunrise to make the descent so that the western part of the crater will be partially illuminated. This will reduce the risk of falls.

a cave near Moltke, consisting of an old lava tunnel that has collapsed

The terminator is about an hour's walk away. Once you reach the terminator, you can watch the Sun rise as many times as you want by walking back and forth between "day" and "night." Don't forget to put the gold visor back on your space helmet each time you step back into "day." The crater is totally illuminated an Earth day and a half after sunrise.

Moltke was named after Count Helmuth Karl Bernhard Moltke (1800-91), a Prussian general who led the 1864 German-Danish War and the 1871 Franco-Prussian War. In addition to his military exploits, he was also a successful cartographer and republished the map of the Moon originally drawn by Julius Schmidt.

Armstrong, Aldrin, and Collins Crater Pits

These three crater pits, named after the crew of the Apollo 11 mission, lie due north of the Moon Museum. The craters range in diameter from 2.4 to 4.5 km, and in depth from 500 to 650 meters.

On lunar maps printed before 1970, Armstrong was called Sabine E. On 21 July 1969, Neil Armstrong, commander of the Gemini 7 and Apollo 11 missions, became the first human to set foot on the Moon. Edwin E. (Buzz) Aldrin Jr., the pilot of the Gemini 7 and Apollo 11 Lunar Lander, was second. Michael Collins co-piloted the Gemini 10 (July 1966) flight and piloted the Apollo 11 Command Module.

Passengers taking the monorail to Lunox will pass very near Collins. According to Lunox wags, the monorail was deliberately designed to go by Collins so that the unfortunate astronaut, who had to continue orbiting while his buddies were walking around on the Moon, would not be forgotten.

Surveyor 5

Surveyor 5 lies about a third of the way between Collins and Aldrin. On 11 September 1967, this unmanned spacecraft landed in Mare Tranquilitatis as a prelude to the first manned mission to the Moon. The camera aboard Surveyor 5 took 19,054 high-resolution photographs.

This lunar spacecraft also analyzed the upper layer of the lunar *mare* soil. It bombarded the atomic nuclei in the lunar soil with radioactive alpha particles, which were reflected back to the Surveyor. Some of the nuclei released their protons, giving off energy. The Surveyor then recorded the energies of the reflected alpha particles and of the protons and used this data to type the atomic nuclei. Based on these experiments, scientists concluded that the upper layer of the lunar soil at that site was very similar to the ordinary basalt found on Earth.

As with the landing gear of Apollo 11, you can examine Surveyor 5 only from a walkway five meters above the site, ever since one of the Moon's very first tourists years ago stole one of its cameras. But since footprints never fade on the Moon, the souvenir hunter was identified by the tread of his moonboots and prosecuted.

LUNOX

From the Moon Museum, you can catch the northern-bound monorail to the ghost crater Lamont, headquarters of the Lunox oxygen and concrete production plant. Lunox is a joint venture set up by the Japanese construction company Shimizu Corporation and the Texas engineering firm Carbotek Development Laboratories, which was conducting tests concerning the feasibility of extracting oxygen and producing concrete from moon rocks as far back as the 1980s.

Lunox's "electromagnetic gun" in Lamont. Canisters of oxygen and concrete are continuously being fired at manufacturing colonies in space.

It appeared from chemical analysis that the moon rocks brought back by the Apollo missions were high in oxygen (41.3%). The remainder consisted of silicon (21.6%), iron (15.3%), calcium (6.9%), magnesium (6.8%), and aluminum (5.4%).

The Taiwanese concrete expert Dr. Tung Dju Lin also tested moon rock samples. He discovered that he could produce a concrete derivative five times stronger than the concrete made on Earth from a compound of iron, titanium, and oxygen in the rock known as ilmenite.

Both Lamont and the area around Mt. Schneckenberg are rich in ilmenite, and the lunar surface at both sites is being excavated. The rocks are pulverized and heated to a temperature of 800° C in reactors. Although excavation continues throughout the night, the actual production process takes place during the day, when sufficient energy can be generated from Lunox's massive solar panels. The oxygen obtained from the rock is stored in cylinders and transported to the various lunar bases along with the concrete produced.

Until a few years ago, Lunox manufactured only enough oxygen and concrete to meet the needs of the Moon. But now that most of the bases have been completed, Lunox has turned into a flourishing export company.

When you arrive at Lunox, the first thing to catch your eye is a huge "electromagnetic gun" that stretches for kilometers over the lunar terrain. This railgun slings canisters of oxygen and concrete to "libration points" L4 and L5 in the Earth-Moon system (60° ahead and 60° behind the Moon in its orbit around the Earth), where manufacturing colonies have been established. These colonies use "mass catchers" to intercept the raw materials, which are then processed into either rocket fuel or construction materials for space stations.

MARE TRANQUILITATIS TO MARE VAPORUM

The monorail links the Lunox headquarters at Lamont, in Mare Tranquilitatis, to the other ilmenite-rich area near Mt. Schneckenberg, at the edge of Mare Vaporum. This north-bound section of the route is 560 km long. The freight train makes the run in a little less than 90 minutes.

During the lunar day a slower monorail for tourists departs every 48 hours. The tourist train makes two stops, each of which lasts 90 minutes: one at the entrance to Ariadaeus Rille and the other at the point where it meets the more southerly Rima Ariadaeus I.

Ariadaeus Rille and Ariadaeus Crater

The monorail stops first at Rima Ariadaeus, otherwise known as Ariadaeus Rille, which connects Mare Tranquilitatis and Mare Vaporum. The nearly straight rille is more than 300 km long and three to five km wide. It is virtually level, except for a small ridge about halfway. A tunnel was built through this ridge in 2015.

Seen from afar, this gorgeous fault region resembles the Taroko Gorge in Taiwan. If you stand on the north rim of the 300 meter high crater wall of Ariadaeus A, you can see for approximately 32 km. To the north are four small crater pits, and behind them the valley that stretches out from Ariadaeus E. To the south you can see the gaping hole of Ariadaeus A (with a diameter of 8.5 km and a depth of 1,000 meters). Beyond that lies the crater Ariadaeus.

THE NEAR SIDE   109

Ariadaeus Rille cuts through the lunar landscape between Mare Tranquilitatis and Mare Vaporum. There is a fantastic view from the plateau above it.

Ariadaeus Crater was named after Philippus III Arrhidaeus (circa 375-317 B.C.), a half-brother of Alexander the Great. Despite his epilepsy, Arrhidaeus was crowned king of Macedonia after Alexander died in 323 B.C., but was put to death five years later by his rival, Olympias. Arrhidaeus supposedly drew up a list predicting solar eclipses. Years later Dionysius (A.D. 9-120) wrote a legendary letter in which he claimed that the darkness that descended on the Earth during the crucifixion of Jesus was not included on Arrhidaeus' list.

Rima Ariadaeus I
Rima Ariadaeus I, just 45 km from Ariadaeus A, is a 30-km-long tributary that runs at a 60° angle into the main Ariadaeus Rille. The view from here is fantastic. An elevator on the north wall of Ariadaeus conducts visitors from the crater floor to an 800-meter-high plateau. From the top you can see ravines running in three directions for more than 50 km. Magnetized overshoes are available for elevator passengers, as otherwise the low gravity would make going up and down very difficult.

After a stop of nearly an hour, the monorail continues through Ariadaeus Crater. Sixty km later, it zooms through the tunnel cut in the ridge, and after another 40 km begins climbing the hills around Silberschlag.

Silberschlag Crater
The main crater Silberschlag and the smaller crater Silberschlag A are 15 km to the left and right of the hills. The main crater, lying off to the south, is 13.5 km in diameter and 2,550 meters deep. Silberschlag A, lying off to the north, is seven km in diameter and 1,800 meters deep.

The craters were named after Johann Silberschlag (1721-91), a German theologian and astronomer. He was one of the first to calculate the trajectory of a meteor.

From Silberschlag the monorail will travel another 160 km across the level Mare Vaporum to Mt. Schneckenberg, which is crawling with miners and excavation equipment.

MT. SCHNECKENBERG

Mount Schneckenberg, also known as Spiral Mountain, is a peculiar, snail-shaped formation with a large dome-top crater in the middle. A small crater sits at its peak. So little sunlight is reflected from the surrounding area that 20th-century scientists thought this might be a volcanic fault area. They were right. The lava deposits that have been found are so rich in ilmenite that Lunox is conducting extensive excavations here. The miners are slowly whittling Mt. Schneckenberg away, which explains why it is no longer included on the list of the Moon's most prominent mountains.

Hyginus Rille
Mount Schneckenberg lies a day-trip away from the selenologically interesting Hyginus Rille (Rima Hyginus), 60 km to the south. The Dutch astronomer and mathematician Christiaan Huygens discovered and described it in the 17th century.

Unlike Ariadaeus Rille, which runs neatly over the slopes of sunken crater walls, Hyginus Rille slices the landscape in two. In the north, it is actually a chain crater, a progression of tiny unwalled craters that join to form one long cleft. Seen from above, it resembles the imprint that a string of pearls might leave in soft clay.

Hyginus Crater
The crater Hyginus, 10.5 km across and 750 meters deep, lies in the middle of Hyginus Rille. Its walls have collapsed in several places, giving it a very weather-beaten look. Hyginus is not an impact crater but a collapsed caldera created by volcanic activity. It has several shallow crater pits, just like the secondary crater pits in Hawaii's Kilauea caldera, and the floor is dotted with little dome-top craters.

The most beautiful view here is the one from the western wall. Looking northward, you can see a chain crater consisting of no fewer than 11 separate craters.

Gaius Julius Hyginus (1st century A.D.) was a Spanish slave freed by Emperor Augustus and appointed head of the Palatine Library. He was a friend of Ovid's and the author of *Poetica Astronomica,* an outstanding description of the constellations and their mythology.

MT. SCHNECKENBERG TO MONT BLANC RESORT

Lunar Modules depart for Mont Blanc Resort from the launching platform near Mt. Schneckenberg during the entire lunar day. During the lunar night, there are no regularly scheduled flights. Travelers must either charter their own LMs or detour to Moon City.

It takes just under 20 minutes to travel the 1,146 km between Mt. Schneckenberg and Mont Blanc Resort. Halfway through the flight the LM reaches an altitude of 285 km and you'll be able to see as far as 934 km. From that altitude, the view is something you'll never forget. On your left, to the rear of the LM, you can just make out the bright ray crater Copernicus in the first rays of the Sun. In front of it, in almost a straight line, lies the lovely Eratosthenes Crater.

To the front of the LM, also on your left, is the dark Mare Imbrium. Its mountain peaks and scarps rise like bright white dots; looming at you in a wide curve (looking counterclockwise) are the Spitzbergen Mountains (1,500 meters), Piton Cape (2,250 meters), Pico Cape (2,400 meters), and the seven peaks of the Teneriffe Range (the highest of which is 2,400 meters).

On your right lie the dark lava plains of Mare Serenitatis. The only bright spots are formed by the rubble around Bessel Crater and the silvery glow around the mysterious Linné Crater.

The Apennines

Looking below, you have a fantastic view of Montes Apenninus, otherwise known as the Apennines. This mountain range, extending more than 1,050 km, is 250 km wide at its broadest part. The Apennines are so impressive that the very first Moon observers drew them on their maps.

They rise gradually on the side of Mare Vaporum and reach their highest peaks and steepest slopes at the edge of Mare Imbrium.

They were formed when large chunks of the Moon's crust rose during the formation of the Imbrium basins. Several of the tall peaks have been given names: Mt. Wolff, Mt. Serao, Mt. Ampère, Mt. Huygens, Mt. Bradley, and Mt. Hadley.

LUNACY

THE ROMANS WORSHIPPED THE MOON GODDESS LUNA, WHO WAS ALSO THE sister of the Sun god Sol. King Servius Tullius (6th century B.C.) allegedly built a temple dedicated to Luna on the Aventine Hill in Rome. The temple was later destroyed by the great fire during the reign of Emperor Nero.

In many cultures for thousands of years, people were convinced that there was a connection between the Moon and a person's mental state. The word "lunatic" comes from Luna's name, and the connection between the two also survives in words such as mooncalf and moonstruck.

The Egyptians believed that insanity could be cured by eating meatballs made from a particular snake, under the light of the Full Moon.

Hippocrates said that moonlight caused nightmares; the Talmud warns against sleeping in it; and Plutarch declared that to do so would drive the sleeper insane.

The medieval physician Paracelsus called the brain the "microcosmic Moon," and said that lunacy grew worse at Full Moon. Until the end of the 19th century, people who committed crimes during a Full Moon were routinely given lighter sentences.

Sir William Blackstone, a prominent 18th-century jurist and philosopher, described a crazy person as someone having "moments of clarity, who is sometimes in charge of his mental capacities and sometimes not, according to the phase of the Moon."

The belief that lunacy was governed by the Moon probably came into existence when people discovered that some forms of mental illness were periodic. Needless to say, some attacks of mental illness are bound to coincide with a Full Moon. Yet if this were the rule, psychiatric hospitals would be bursting every time there was a Full Moon.

Mt. Huygens
Keep your eye on the mightiest peak in the Apennines: Mt. Huygens. At 5,400 meters, it towers above the other mountains in this range. It was named after Christiaan Huygens (1629-95), Holland's most renowned mathematician, physicist, and astronomer. Huygens' wave theory of light was later confirmed by other scientists, while the light-particle theory of his contemporary Isaac Newton was rejected.

Incidentally, Newton's name was given to the Moon's deepest crater. The floor of Newton Crater, on the near side of the Moon near the Mountain of Eternal Light, is 8,850 meters lower than its rim.

Plato Crater
Moon flights go so quickly that it's almost time to land. You can see the 3,600-meter-tall Mont Blanc in front of you. Below and to the right is the famous Alpine Valley, which divides the Alps into two approximately equal parts. On the far left lies one of the Moon's most beautiful craters, the flat crater Plato. At sunrise the soil appears gray. When the Sun is halfway to its zenith, the soil turns yellow. When the Sun reaches its zenith, the soil changes to dark black. Of course this is an optical illusion, since Plato must be brightest when the Sun is highest. It only appears less bright because the surrounding highlands have become even brighter.

MONT BLANC RESORT

Mont Blanc Resort is more modern and offers more luxury than Moon City. Operators have made it as pleasant as possible for tourists and workers on leave. The resort was built in 2015 by the Ohbayashi Corporation, a major Japanese construction company.

In the middle of the resort is a 540-meter-tower containing several restaurants, which you can reach by a glass elevator. From the top, you have a magnificent view of Mont Blanc off to the south. The Alpine Valley begins on the northeastern horizon.

The large glass dome at the foot of the tower has an artificial atmosphere and a fertile soil. In the beautiful Christa McAuliffe Memorial Park created inside the dome, the plants and bushes have shot up "to the sky," thanks to the Moon's low gravity. It also serves as a sports and recreational center. You can play lunar golf here, or visit the swimming pool. (See "Sports" in the "Recreation" section of the On the Surface chapter for full details.)

VICINITY OF MONT BLANC RESORT

Mont Blanc
You need no special mountain-climbing equipment to scale Mont Blanc for the simple reason that there are no towering peaks in the Alps. The base of Mont Blanc measures 30 km. The average slope from the access road to the top of the 3,600-meter peak is therefore less than 14°, which means that you can drive a moon buggy the whole 15 km to the summit.

On top of the peak is a depression where you can walk around. From the walls you can see as far away as 112 km. If you look to the north and northeast, you have a view of most of Alpine Valley. To the south, the Deville, Agassiz, and Piton Capes jut out above the horizon.

Alpine Valley
The 190-km-long Alpine Valley is a fault region that was observed for the first time in 1790 by the German lawyer and astronomer Johann Schröter. It is narrowest (two km) at Mare Frigoris. It starts to widen in the southwest (to a maximum of 10 km), but as it nears the Mont Blanc Resort it runs up against a three-km-high mountain and slices it down the middle. After enjoying the magnificent view from the highest point above the cleft, you can climb down to the foot of this mountain, where Alpine Valley forms a kind of amphitheater.

As of this writing, there is no monorail in Alpine Valley. This is unfortunate, as the valley is too long to traverse without one. Still, among the sights in the vicinity of the amphitheater is a rille that meanders through the valley floor. The American astronomer William Henry Pickering, whose name you have encountered before (see "The Origin of the Moon" and "Bugs and Bombs" in the Introduction and "Linné Crater" earlier in this chapter), discovered the rille in 1891.

Alpine Valley seen from above. A rille runs through the middle of the 190-km-long valley and ends in a bowl-shaped "amphitheater." The isolated peak in Mare Imbrium (top right) is Pico.

Plato Crater Recreation Area

The Huygens, the mammoth Lunar Module that operates out of the Mont Blanc Resort, makes the short 330-km hop to Plato in less than 10 minutes. The capes and scarps along the northeastern curve of Mare Imbrium are clearly visible, and Plato also demands your full attention. At sunrise, watching the eastern wall cast its sharp shadows and seeing them progressively shorten is a sight you're not likely to forget.

The Huygens is equipped with more than 48 hours of power and oxygen for the PLSSs. The food in the restaurant on board is decent by flight standards. Those wanting to rest may snooze in the LM's reclining seats. Those in search of adventure will soon discover that Plato, like Mont Blanc Resort, is devoted to sports and recreation. Golfers can try their luck at the Shepherd Driving Range, and landlubbers can indulge in solar sailing. (See "Sports" in the "Recreation" section of the On the Surface chapter for full details.)

MONT BLANC RESORT TO SCHRÖTER'S VALLEY

It takes the Lunar Module about 20 minutes to cover the 1,312 km between Mont Blanc Resort and Schröter's Valley; along the route it reaches a maximum altitude of 326 km.

The flight crosses almost diagonally over Mare Imbrium. Provided you stick to the travel itinerary outlined here, the sea will be in full sunlight. To your right you can see the lovely Bay of Rainbows, enclosed by the Jura Mountains. To your left you can see the Carpathian Mountains, and behind them the gorgeous ray crater Copernicus.

Try to keep your eyes on the landscape during the last five minutes of the flight. To the north of Gruithuisen Crater, named after Franz von Paula Gruithuisen (see "Visionaries and Dreamers" under the "History and People" section of the Introduction, or visit the "Life on the Moon" exhibit in the Moon Museum), you can see several big (roughly 1,800 km tall) dome-top craters. These lunar equivalents of shield volcanoes are responsible for some of the *mare* lavas.

Schröter's Valley was highlighted earlier in this chapter under "Attractions." From here, the Lunar Module will fly you to Moon City.

MOON CITY

The flight from Schröter's Valley to Moon City takes a little less than 20 minutes. Halfway through the trip the LM flies over Hortensius Crater, named after the Dutch mathematician and astronomer Martin van den Hove (1605-39). Better known by his Latin name Hortensius, he wrote about Venus and a transit of Mercury in front of the Sun.

Hortensius Crater lies approximately halfway between two ray craters, Kepler and Copernicus. Lansberg Crater, named after the Flemish doctor and astronomer Philippe van Lansberg (1561-1632), is not far from Moon City.

Moon City is invariably a letdown to those who just spent a week or more in the luxurious Mont Blanc Resort. It is sturdy but dull. Most of the complex was built underground, but the exposed parts are covered by thick slabs of concrete. This was initially necessary to protect the buildings

the industrial park southeast of Moon City

and inhabitants from the harmful rays of the Sun. Six years later Ohbayashi Corporation managed to produce an effective radiation shield from volcanic glass that had been mined on the Moon.

Despite its meager population of 2,400, Moon City possesses the vitality of a true metropolis. Most of the residents are Japanese, which is not surprising since Shimizu Corporation founded the city in 2009 after Lunox went into full operation.

VICINITY OF MOON CITY

All the historic landing and impact sites in and around Moon City will keep interested visitors busy for quite a while. For example, the fully intact landing gears of the Apollo 12 (1969) and Apollo 14 (1971) Lunar Landers lie within walking distance.

The unmanned Surveyor 3 (1967) lies a stone's throw away from the Apollo 12. Some of the parts are missing—it seems the Apollo 12 crew took them back to Earth so scientists could investigate the effects of a long-term stay on the Moon. They discovered a bacteria in the camera of Surveyor 3 that had survived three years on the Moon!

THE REAL MOON CITY

IT'S EASIER THAN YOU THINK TO TRAVEL TO THE MOON. MOON, KENTUCKY, that is, a small town on KY 172. Moon is so tiny (pop. 75) that if you miss the post office, you won't know you passed through town. Someone is always stealing the Moon highway signs. The town did get a little excitement in 1969, when the first astronauts landed on the other Moon. People wanted letters and cards with Moon postmarks. And for a while some folks would come from out of town, collect dirt, and sell it as "Moon dust." In 1989, to commemorate the lunar landing, a Kiwanis Club in a nearby town decided to raise money by selling chances to win a Moon trip. The lucky winner also received a case of Moon Pies and a Moon rock, which was found in a ditch outside of town.

Lying in a wide arc around Moon City are the fragments of various crafts that crashed on the Moon. This is where Surveyor 2 (1966) and Luna 5 (1965) met their untimely ends. And the lift-off stages of the Apollo 12 and Apollo 14 rockets as well as the third and last stages of the Apollo 12-Apollo 17 Saturn V rockets were deliberately crashed in this area. Scientists wanted to study the effects of artificially created moonquakes, which were recorded by the seismometers left by Apollo 12 and Apollo 14. They discovered that the Moon vibrates a long time after every quake.

They made their most effective seismic recording during the Apollo 14 mission. Early on 4 February 1971 the third stage of its Saturn V launcher hit the Moon 210 km southwest of the Apollo 12 site with the energy of 11 tons of TNT. The seismic waves generated by the impact caused the whole region to reverberate for three hours.

MOON CITY TO THE STRAIGHT WALL AND MOUNTAIN OF ETERNAL LIGHT

The famous Straight Wall is a mere 742 km from Moon City. It lies directly on the route to the ice region in the South Pole, near the Mountain of Eternal Light. This spectacular route is a perfect way to round off your trip to the Moon.

The flight from Moon City to the Straight Wall is just a short jump, yet it affords you a fantastic view of the immense Ptolemaeus, Alphonsus, and Albategnius walled plains and the Arzachel ring plain.

After a two-hour stop at the Straight Wall, your route takes you to the Amundsen walled plain, 2,075 km away. This is too far for the Lunar Module's normal energy-saving hop. Although the LM seems to rise to a normal altitude of 50 km, it quickly gathers speed and follows a horizontal trajectory at a near orbital velocity. Twenty minutes later the LM is racing over the southern highlands, past the magnificent Tycho Crater, over the immense Clavius walled plain and over Newton, the Moon's deepest crater. You'll also see Moretus and Malapert, two craters named after Belgian mathematicians.

Then the LM begins its descent into the dark Amundsen Valley, whose only point of light is the tip of the Mountain of Eternal Light. A laser beam, which you can see flickering on and off in the distance, guides the LM with utmost precision to the landing platform. Since up to now you've been flying exclusively in daylight, you haven't had the opportunity to notice the laser-guided navigation system before. But if you look out the window of the LM, you can clearly see a dark-red light, gleaming like a necklace of sparkling rubies in the inky blackness of the night.

THE FAR SIDE

The far side of the Moon doesn't yet have overnight accommodations for tourists, so you'll have to make a day-trip. The far side holds no secrets, except from Earthbound humans, who never see its face. Lunar travelers can visit easily during specified times and will find it an eye-opening trip.

Radio Silence
In the early days of space travel, when spacecrafts swung behind the Moon to come into full orbit, mission control on Earth had to hold its breath during the ensuing period of radio silence. The Moon blocked any radio signal to and from the Earth, and astronauts on the far side were on their own. This proved critical during the Apollo 13 mission (1970), when

the spacecraft had to be guided back to Earth after an explosion blew up part of its service module. Now, thanks to continuous-relay communication satellites in orbit around the Moon, these long minutes of radio silence are a thing of the past.

Even so, the deserted character of the far side has much to do with radio silence. Because the far side of the Moon is always turned away from Earth, any noise—radio, TV, or strong radar signals—is blocked by the Moon. No wonder astronomers long eyed the far side of the Moon as a prime site for a large radio and optical observatory that would let them look into the galaxy. They finally got their wish: in 2009, while Moon City was being built on the near side, astronomers and technicians were swarming over the far side, building their long-dreamt-of observatory.

EYE ON THE STARS

The Lunar Far Side Observatory

While astronomers perform their galactic observations, radio silence must be observed for travel between the near and far sides of the Moon. Without beacons and relays to communications satellites, travel is unsafe

---

MOONBOUNCE

IN RECENT DECADES THE MOON HAS BEEN USED FOR COMMUNICATIONS AS A passive radio signal reflector. The technique is known as Earth-Moon-Earth (EME) or Moonbounce. High-powered signals are transmitted from antennas pointed at the Moon. The Moon acts like a giant radio mirror in the sky, bouncing the radio waves back to listeners on Earth.

The first radar signals to echo from the Moon were transmitted on 10 January 1946, by John H. DeWitt Jr. as he was working on the Pentagon's Project Diana, named for a mythical Moon goddess. The project scientists were trying to find out if missiles could be detected above Earth's ionosphere.

The first ham radio signals to echo from the Moon were transmitted in 1953. Today, radio amateurs regularly beam signals to the Moon, with their reflections mapping a large area of our planet.

and limited to emergencies. But for one week each month the observatory is fully lit by the Sun, whose noise and fierce light interferes with both radio and optical observations. Lunarians and tourists alike seize the opportunity and flock to Tsiolkovsky crater to catch a glimpse of the now-famous Lunar Far Side Observatory (LFSO).

You can only gaze in awe at the tremendous radio dish that fills a secondary shallow crater inside Tsiolkovsky crater. It is an "in-ground radio dish," very much like the terrestrial radio observatory in an almost circular valley near Arecibo, Puerto Rico. Arecibo Observatory measures a respectable 300 meters in diameter. But LFSO's radio dish antenna dwarfs it with a diameter of one and a half kilometers!

The dish antenna maps the gas in the Milky Way and other galaxies and also canvasses the neighborhood for signs of extraterrestrial intelligence. The in-ground radio dish may be LFSO's largest single instrument, but LFSO's radio interferometer performs the real study of the skies. Its 27 dishes, each 50 meters wide, form a Y with 20-km arms to simulate the aperture of one huge disk. By using this radio synthesis telescope to examine distant galactic clusters, lunar astronomers hope to be able to determine the universe's expansion rate one day.

The southeastern part of 193-km-wide Tsiolkovsky crater houses LOUISA, the Lunar Optical Ultraviolet-Infrared Synthesis Array. LOUISA consists of 42 telescopes arranged in two circles. In 2018 this multipurpose array found dim planets around a multitude of other stars, which finally proved that our solar system is not unique and that rocky planets like our Earth are a common aspect of star formation.

The Moon's lack of atmosphere means that astronomers can set their X-ray detectors on the surface. So not far from LOUISA you'll find LAMAR, the Large Area Modular Array of Reflectors. LAMAR is a matrix of X-ray collectors spread out over an area of 30 square meters. This adds up to an X-ray collection area of up to a million square centimeters. That's 1,000 times more sensitive than AXAF, the Advanced X-ray Astrophysics Facility that the U.S. space agency NASA put into orbit during the last decade of the 20th century.

Chain of Telescopes

The Lunar Far Side Observatory has even more surprises in store. A visit to its data-collection/processing center reveals that radio astronomers have electronically linked LFSO's radio interferometer to the Very Large Array (VLA) of radio telescopes in New Mexico—creating a 384,000-km-wide instrument that can measure the locations of radio astronomical features up to an accuracy of 30 millionths of a second of arc. This Moon-Earth Radio Interferometer (MERI) is tackling a long list of astrophysical and cosmological problems. Among other research subjects, it is recording for the first time ever the radio emissions of individual stars such as the Sun, even though they are as remote as 1,000 light-years!

The Tour

A complete tour of the observatory takes a full seven Earth hours, including lunch at the data-collection/processing center. Since the overnight facilities are limited to astronomers and technical personnel, you won't arrive back at your hotel on the near side until long after dinnertime. But don't pass up the opportunity to visit the observatory because you might miss a meal! It's more than worth it, and members of the astronomical staff are happy to share their discoveries with visitors.

THE MOON ILLUSION

Nearly everyone who looks at the Moon thinks that it's larger and appears closer when it's just come up. Especially during Full Moon, many people think that the rising or setting Moon looks like a huge round ball or balloon. But this is an exaggeration. You can check this out yourself by holding a pencil in front of your eye at arm's length. You will find that the Moon's disk will always be covered by the pencil, no matter whether the Moon is on the horizon or high in the sky.

Photographs have shown that we are dealing with a peculiar kind of optical illusion. When measured on a negative, a Moon photographed low on the horizon is exactly the same size as one photographed high in the sky. But then something strange happens. If you print the photographs and look at them again, the rising or setting Moon looks a lot bigger.

Since antiquity, people have been trying to explain this "Moon illusion." Of the variety of explanations that have been proposed, two are worth considering here: the so-called apparent-distance theory and the angle-of-view theory. According to the apparent-distance theory, the rising or setting Moon seems larger because the Moon appears to be farther away. According to the angle-of-view theory, the Moon high in the sky seems smaller because the observers have to move their heads or eyes upward to see it.

The Apparent-Distance Theory
The apparent-distance theory was described long ago by the Greek astronomer and geographer Claudius Ptolemaeus, better known as Ptolemy (A.D. 87-150). According to Ptolemy, an object that is viewed above a landscape will give the impression of being farther away than an object that is equally distant but that is viewed in the middle of an empty space, such as the Moon high in the sky. Although the image seen by the eye is the same size in both cases, the object that seems farther away looks larger.

The Angle-of-View Theory
The angle-of-view theory dates from the early 1950s and tries to take into account the elevation of the Moon above the horizon. People move their heads and eyes as they look at a Moon high in the sky and the theory states that this change in position would account for the illusion.

(continues)

When the Moon has just risen it seems huge, especially when there's a clear view of the horizon. The effect is the strongest when the horizon seems far away.

However, in an experiment carried out in 1962 at the Hayden Planetarium in New York, American psychologists Lloyd Kaufman and Irvin Rock used projected images of artificial Moons to effectively prove that the Moon illusion cannot be caused by a change in the observer's position. Subjects who were asked to observe two identical Moons, first with their eyes raised and then with their eyes not raised, did not report any significant difference in size.

Kaufman and Rock therefore carried their experiment a step further. They used a sheet of cardboard with a hole in it, which effectively shielded the Moon from the rest of the sky. They found that the subjects estimated that the Moon rising on the horizon was 34% larger than the same Moon with the horizon covered by the cardboard. This could only mean that Ptolemy was right, and that the horizon is responsible for the peculiar illusion.

As far as we can tell now, our brain's perception of distance creates the Moon illusion. The sky above us is infinite. And yet we don't perceive it this way; we think of it as flattened. So the sky above us seems near

When the Moon is high in an empty sky, it appears to be very small. And yet this Moon has the same diameter as the one in the previous picture.

and the horizon seems far away. Though our eyes truthfully register two images that are the same size, our brains trick us into thinking one must be larger than the other because it appears to be farther away.

In reality, the diameter of the Moon is always the same, no matter where we see it. But wait. The Moon is not always at the same distance from the Earth. The average distance between the Moon and the Earth is 384,000 km. But this is merely an average—the Moon doesn't rotate in a circle but in an ellipse, so that the distance between the two bodies varies from 356,375 km to 406,720 km. This means that the apparent diameter of the Moon can be as much as 14% larger at its nearest point from the Earth compared to its farthest point from the Earth.

But this is just the astronomical background to the story. To compare these sizes of the Moon's disk, you'd have to observe the Moon for a long time at calculated intervals. And who can remember how large the Moon was a few weeks ago? This effect can only be seen on a series of photographs taken through a telescope.

AFTERWORD: EARTH WITHOUT THE MOON

What would the Earth look like without the Moon? Earth and Moon make up a unique double planet. And it may very well be that our giant Moon was necessary for the emergence of life.

The Moon is far larger compared with the Earth than any other planet/satellite pair in our solar system. Our Moon is so large that it does not actually go around the Earth at all. That is, not around the center of the Earth. Earth and Moon wobble around a common point of gravity, the barycenter. When the Moon is directly overhead, the barycenter is only about 1,713 km (1,000 miles) beneath our feet.

The ocean tides caused mainly by the Moon (the Sun has half as much tidal effect on us as the Moon) had a profound effect on the evolution of crustaceans and amphibians. The existence of intertidal zones—the regions exposed on the shoreline at low tide and covered at high tide—perhaps even helped life to emerge on land.

The ocean tides on Earth are not a fraction of what they once were. About 4.5 billion years ago, after an object the size of a small planet struck primeval Earth, the Moon formed from the resulting ring of debris that circled our planet. For a billion years thereafter, the Moon was still very close to the Earth. This proximity created enormous tides on both bodies and heated their cores far above what would be normal. This prolonged heating continued until the action of the tides pushed the Moon to nearly its present distance and slowed the rotation of the Earth from about once every four hours to once every 24 hours.

Some of the Earth's unusual characteristics may be due to the prolonged heating of its core. Unlike Venus or Mars, the Earth still has an active, molten core. The energy of this core drives the tectonic plates, recycles new crust from the interior, and releases gases from inside the Earth. Thus, the molten core is responsible for all our volcanoes and mountain ranges and for the separation of the continents, which has isolated gene pools and speeded up evolution.

The Earth also possesses a very powerful magnetic field. It extends far into space and protects life from deadly cosmic radiation. This field is a hundred times stronger in proportion to our planet's mass and angular momentum than that of any other planet. Scientists believe that this exceptionally strong field also is related to our still active, molten core.

So without the Moon life may never have flourished on Earth. Large double planets such as ours are very rare. And that means that humans may be the only, or one of the very few, civilizations in our galaxy.

"Where are the extraterrestrials? Why haven't they landed in their flying saucers on the White House lawn to welcome humanity to the Galactic Club?" Nuclear physicist Enrico Fermi asked this question back in 1939, long before our search for extraterrestrial intelligence began.

Fermi's paradox rises from a chain of apparently sound logic. The Milky Way is a pinwheel disk of about 200 billion stars. Many of these may be accompanied by planets, and thousands or even millions of these planets could conceivably support life. If life arises naturally, then we might expect thousands of planets crawling with life and perhaps hundreds of civilizations in our galaxy. Yet there are no signs that these civilizations have visited us. Nor have our radio telescopes picked up any extraterrestrial communications.

Thus life on Earth may be unique. And we may be the first civilization to explore and colonize the barren worlds in our galaxy. With the Moon, we have taken the first step. And although we still have very few settlements there, soon the Moon will support a whole population of Lunarians—the first generation of humans accustomed to living and working in space. They are the proof that we can carry on further.

Imagine 50 years from now. The small crescent of the Moon becomes visible after sunset. There, on the darkened surface of the nighttime side, shining like stars in places where no stars ought to be, we see the sparkling lights of lunar cities. They are the lights of humanity's second planetary civilization.

RECOMMENDED READING

Brueton, Diana. *Many Moons.* New York: Prentice Hall Press, 1991.

Burgess, Eric. *Outpost on Apollo's Moon.* New York: Columbia University Press, 1993.

Burns, Jack O., Nebosja Duric, G. Jeffrey Taylor, and Stewart W. Johnson. "Observatories on the Moon." *Scientific American,* March 1990.

Burrows, William E. "Securing the High Ground." *Air&Space,* Dec. 1993-Jan. 1994.

Cain, Kathleen. *Luna, Myth, & Mystery.* Boulder, CO: 1993.

Chaikin, Andrew. *A Man on the Moon: The Voyages of the Apollo Astronauts.* New York: Viking Penguin, 1994.

Chaikin, Andrew. "Shoot for the Moon." *Air&Space,* Dec. 1991-Jan. 1992.

Compton, William David. *Where No Man Has Gone Before.* Washington, D.C.: NASA, 1989.

Cordell, Bruce. "Search for the Lost Lunar Lakes." *Astronomy,* March 1993.

Cortright, Edgar M. *Apollo Expeditions to the Moon.* Washington, D.C.: NASA, 1975.

Evans, Davis S. "The Great Moon Hoax." *Sky & Telescope,* Sept.-Oct. 1981.

Heiken, Grant H., David T. Vaniman, and Bevan M. French. *Lunar Sourcebook, A User's Guide to the Moon.* Cambridge, 1991.

Hurt, Harry. *For All Mankind.* New York: Atlantic Monthly Press, 1988.

Kuznik, Frank. "Personal Effects." *Air&Space,* December 1994.

Lewis, H.A.G. *The Times Atlas of the Moon.* London: Times Newspapers, 1969.

Ley, Willy. *Watchers of the Skies.* London: Sidgwick & Jackson Ltd., 1963.

Moons and Rings. Alexandria, VA: Time Life Books, 1991.

Moore, Patrick and Iain Nicolson. *The Universe.* Oxford: Equinox Ltd., 1985.

Moxon, Julian. "Return to the Moon." *Flight International,* 15 July 1989.

Shepard, Alan and Deke Slayton. *Moon Shot. The Inside Story of America's Race to the Moon.* Atlanta: Turner Publishing, Inc., 1994.

Short, Nicholas M. *Planetary Geology.* Englewood Cliffs, N.J.: Prentice Hall, 1975.

Wilhelm, Don E. *To a Rocky Moon. A Geologist's History of Lunar Exploration.* Tucson: University of Arizona Press, 1993.

INDEX

Italicized page numbers indicate information in captions, charts, illustrations, maps, or special topics.

A
Abian, Alexander: *23*
accommodations: 76
Advanced X-ray Astrophysics Facility (AXAF): 121
Aldrin Crater Pit: 105-106
Aldrin, Edwin E. (Buzz) Jr.: 38, 60-61, *61*, 105-106
Alphonsus Crater: 98-99
Alpine Valley: *28*, 34, 114, *115*
Alter, Dinsmore: 98
aluminum: 36
American Moon missions: 57-61, *58, 59, 60, 61*
Amundsen Valley: 103-104
Amundsen Walled Plain: 119
angle-of-view theory: *123-124*
animals associated with the Moon: *52*
Apennines: 29, 111-113
Apollo program: 25, 60-61, *60, 61*
Apollo spacecraft: 71, 117-118, 119-120
apparent-distance theory: *123*
area: 13
Arecibo Observatory: 121
Ariadaeus Crater: 108, *109*
Ariadaeus Rille: 34, 108, *109*
Aristarchus Crater: 22, 100
Armalcolite: 35
Armstrong Crater Pit: 105-106
Armstrong, Neil: 38, 60-61, *61*, 105
Arrhidaeus, Philippus III: 109
astrotrash: 38-40
attractions on the Moon: 97-104, *98, 102, 103*
AXAF: 121

B
bacteria: 51-52
Bailly walled plain: 32

barycenter: 126
Bessel Crater: 111
Blackstone, Sir William: *112*

C
Caesar, Julius: *91*
calcium: 36
capes: *30-31*
Christa McAuliffe Memorial Park: 113
Chronicle of Gervase of Canterbury: 23
Clarke, Arthur C.: 10
Clavius walled plain: *29*, 32, *38*, 119
climate: 36-37; Moon's effect on Earth's *19-20;* temperature 36; travel seasons 36-37; without the Moon 23
clothing: 89
Collins Crater Pit: 105-106
Collins, Michael: 60-61, *61*
communications: 88
concrete production: 106-108
Copernicus Crater: 22, 32, 111
crater cones: 34
craters: 32-34; Alphonsus 98-99; Ariadaeus 108, *109;* Aristarchus 22; Bailly 32; Bessel 111; Clavius 32; Copernicus 22, 32, 111; crater cones 34; dome top craters 34; Eratosthenes 111; Giordano Bruno 23, *24-25*, *25;* Hortensius 116; Hyginus 110-111; Langrenus *33;* Lansberg 116; Linné 101, 111; Malapert 119; Moltke 104-105; Moretus 119; Newton 113, 119; Plato *28*, 113; ray craters 32; ring plains 32; Silberschlag 109-110; Tsiolkovsky 121; Tycho 22, *29*, 32, 119; walled plains 32

D
Darwin, Sir George Howard: 13, 18, 21
days, length of: 22
de Bergerac, Cyrano: 44-45, *45*
de la Rue, Warren: *46*
distance between Earth and Moon: 21-22, *21*
dome-top craters: 34

E
Earth: distance between Earth and Moon 21-22, *21;* Moon's influence on *19-20;* rotation 21-22; without the Moon 126
Easter: *91-92*

eclipses: 72-75, *73, 75*
eels: *20*
electromagnetic gun: 108
environmental issues: 38-40
Epsilon Peak: 29, 103
Eratosthenes Crater: 111
events: 72-76; eclipses 72-75, *73, 75*; moonquakes 74-76
exchange rate: 88
features: *30-31, 35*; names of *14-17, 27*; *see also specific feature*
Fermi, Enrico: 127
first human-made object on the Moon: 56
first manned landing on the Moon: 60
first manned mission to orbit the Moon: 60
first words spoken on the Moon: 61
fission theory: 13, 18, 21
Flammarion, Camille: 102-103
food: 77-79
From the Earth to the Moon: 79, 80

G
Galilei, Galileo: *43*
geocosmic influences: *19-20*
geography: 25-36, *27, 30-31, 35*; craters 32-34; mountains 29-32; rocks 34-36; seas 26-28, *see also maria*; valleys, rilles, and scarps 34-35
geology: 22-25; *see also* mining
ghost craters: 28
Giambattista Riccioli: 25
giant-impact theory: 22
Giordano Bruno: 25
Giordano Bruno Crater: 23, *24-25*, 25
Goddard, Dr. Robert Hutchings: 3
golf: 40, 70-71, 114
Gonsales, Domingo: 43-44
Goodwin, Francis: 43-44
government: 65-66
Grant, Dr. Andrew: 48
gravity: 93, *93*
Greenwich mean time: 92
Gruithuisen, Franz von Paula: 45-46, *47*

H

ham radio: *120*
harvest moon: 92
health and safety: 85-87
Herschel, Sir John: 47-51
high jump: 69
Hippocrates: *112*
history: 41-63; modern-day travelers 53-63; visionaries and dreamers 41-53
Hiten Satellite: 63
hoaxes: 47-51, *49, 50*
Hortensius: 116
Hortensius Crater: 116
Hu, Wan: 53
Huygens, Christiaan: 110, 113
Hyginus Crater: 110-111
Hyginus, Gaius Julius: 111
Hyginus Rille: 110

I

ice: 64-65, 103-104
ilmenite: 35, 107, 110
impact craters: 22
impacts: asteroids 28; comets 28; creating craters 33-34; meteorites 23; spacecraft 54-63
information and services: communications 88; health and safety 85-87; inoculations 85; officialdom 85; money 87-88, *87;* tourist information 88; weights and measures 89-93, *90-92, 93;* what to take 88-89
International Astronomical Union: 25, *56*
International Lunar Quinquennium: 65
Interplanetary Flight: 10
iron: 35, 36
Jack and Jill: *40*
Jansen, Zacharias: 43
Japanese Moon missions: 63

K

Kaufman, Lloyd: *124*
Kennedy, John F.: 54
Kepler, Johannes: 41-43
Kozyrev, Nikolay A.: 98-99

L

LAMAR: 121

land: 13-40, *14-17, 19-20;* climate 36-37; environmental issues 38, 40; origin 13, 18, 21-22; geography 25-36; geology 22-25; visibility 38, *39*

Langrenus walled plain: *33*

Lansberg Crater: 116

Large Area Modular Array of Reflectors (LAMAR): 121

legends: *19-20, 26, 37, 40, 44, 52, 74, 99, 112*

libration points: 108

Lin, Dr Tung Dju: 107

Linnaeus: 101

Linné Crater: 101, 111

Linné, Karl von: 101

Lippershey: 43

Lipsky Plain: 56

Locke, Richard Adams: 47-51, *49*

LOUISA: 121

Lucian of Samosata: 41

Luna spacecrafts: 28, 54, *54-55, 56,* 118

lunacy: *112*

Lunar Age: 63-66

Lunar Archives: *64*

Lunar calendar: *90-92*

Lunar Far Side Observatory: 120

Lunar Optical Ultraviolet-Infrared Synthesis Array (LOUISA): 121

Lunar Orbiter Program: 59, *60*

Lunar Roving Vehicle: 61

Lunox: 106-108, *107*

M

magnesium: 36

magnetic field: 127

Malapert Crater: 119

Mare Crisium: 28

Mare Imbrium: 28, *28,* 111-112

Mare Moscoviense: 28

Mare Serenitatis: 28, 111

maria: 13, 22, 26-28

marine animals: Earth's influence on *20*

menstrual cycle: 18, 20

MERI: 122

meteorite strike: 23
minerals: 35-36
mining: 106-108, 110; ice 103-104
Moltke, Count Helmuth Karl Bernhard: 105
Moltke Crater: 104-105
money: 87-88, *87*
Mont Blanc: 113, 114
Mont Blanc Resort: 113-114; *see also* accommodations
Montes Sovietici (Soviet Mountains): *56*
Moon bases: *68*
moonbounce: *120*
Moon cakes: 78
Moon City: 116-117; vicinity of 117-119
Moon creatures: 42-43, 45-46, 47-50, *49, 50* 51-53
moondust: 86-87, *86*
Moon-Earth Radio Interferometer (MERI): 122
Moon illusion: *123-125*
Moon, Kentucky: 118
Moon landings: *60;* first manned 60
Moon missions: American 57-61, *58, 59, 60, 61;* Japanese 63; Russian 54-57, *54-55, 56*
Moon museum: 94-97, *96*
Moon pies: 77
moonquakes: 13, 74-76, 118
moonsuits: 88-89
Moretus Crater: 119
Moscow Sea: 28
mountain climbing: 114
Mountain of Eternal Light: 32, 102-104, 110; 113; 119
mountains: 29-32, *30-31*
Mt. Huygens: 29, 113

N

names of features: *14-17, 27; see also specific feature*
Near Side, the: 104-119, *105, 107, 109, 115, 117, 118*
Nero, Emperor: 112
Newton Crater: 113, 119
nomenclature: 25
O'Neill Bridge: 101-102, *102*
O'Neill, John J.: 101-102

O

Ocampo, Sixto: 51
Oceanus Procellarum: 28
officialdom: 85
olivine: 35
origin of the Moon: 13, 18, 21-22
ownership of the Moon: 65-66
oxygen: 35
oysters: 20

P

packing: 88-89
palolo worm: 20, 23
Parcelsus: *112*
people: 41-63; modern-day travelers 53-63; Moon's influence on *20*; visionaries and dreamers 41-53; *see also specific person*
Pickering, William Henry: 51, 114
Pico Cape: 111
Piton Cape: 111
plagioclase: 35-36
planning your trip: 67
Plato Crater: *28*, 113
Plato Crater Recreation Area: 115
Plutarch: *112*
Pope Gregory XIII: *91*
Portable Life Support System (PLSS): 37
potassium: 35, 36
precipitation: Moon's effect on Earth's *19-20*
Ptolemy: *123*
pyroxene: 35
pyroxferroite: 35

R

radiation: 86
radio signal reflector: *120*
radio silence: 119
Ramadan: *91*
Ranger Program: 57, 58
recreation: 69-76, *70, 73, 75*; events 72-76; sports 69-72
regolith layer: 13
rilles: 34, *35*

rima: 34, 35
Rima Ariadaeus 1: 109
Rock, Irvin: *124*
rocks: 34-36
Rook Mountains: 29
rotation: 21-22
rupes: 34, 35
Rupes Recta: *see* Straight Wall
Russian Lunar Rovers: 55
Russian Moon missions: 54-57, *54-55, 56*

S
safety: 85-87
scarps: 34, 35
Schmidt, Julius: 10, 101
Schröter Crater: 45, *47*
Schröter, Johann: 100, 114
Schröter's Valley: 34, 100
seas: 13, 22, 26, 28
seasons: 36-37
Selene: *26*
selenes: 87-88
Selenites: *see* Moon creatures
Selenotopographische Fragmente: 100
Shepard, Alan: 40, 71
shot put: 69
Sidereus Nuncius (The Starry Messenger): 43
sightseeing highlights: 67-69
Silberschlag Crater: 109-110
Silberschlag, Johann: 110
sodium: 35
soil samples: 55, 57, 61
solar sailing 72
Somnium ("The Dream"): 41-43
Soviet Mountains (Montes Sovietici): *56*
space law: 66
spacecraft: 53-63, *54-55, 56, 58, 59, 60,* 94, 117-118; Apollo 60-61, *60, 61;* crashed remains of 38, 40; first launch 53; Hiten satellite 63; Lunokhod 55, 57; Luna 28, 54, *54-55, 56;* Lunar Roving Vehicle 61; Rangers 57, *58,* 98; Russian 54-57, *54-55, 56;* Surveyors 57-59, *58, 59;* Zond *54-55; see also specific spacecraft*

space race: 53-63, *54-55*, *56*, *58*, *59*, *60*, *61*, *62*
spinel: 35
Spiral Mountain: 110
Spitzbergen Mountains: 111
sports: 69-72; golf 70-71; high jump 69; mountain climbing 114; shot put 69; solar sailing 72; swimming 71-72; track and field 69-70
stereopticon: 46
Straight Wall: *29*, 34, 97-98, *98*, 118
surface features: 13
Surveyor program: 57-59, *58*, *59*
Surveyor spacecraft: 106, 117-118
swimming: 71-72, 114

T
Talmud: *112*
telescopes: 43
temperature: 36
Teneriffe Range: 111
The Man In The Moon: Or A Discourse Of A Voyage Thither: 43-44
The Sun: 47-51
thorium: 36
tides: *19-20*, 126
time: 89-92, *90-92*; length of days 22
titanium: 35, 36
tourist information: 88
track and field: 60-70
transportation: 79-85; *80*, *83*, *84*; getting around 84-85, *84*; getting there 79-84, *80*, *83*;
Tsiolkovsky Crater: 121
Tsiolkovsky, Konstantin Eduardovich: 53
Tullius, King Servius: *112*
Tycho Crater: 22, *29*, 32, 119

UV
uranium: 36
V-2 missiles: 54
vallis: 34, *35*
valleys: 34, *35*
van den Hove, Martin: 116
van Langren, Michiel Florentius: *33*
van Lansberg, Phillippe: 116

Verne, Jules: 79, *80*
Virgin Mary: 26
visibility on the Moon: 38, *39*, 82, *82*
von Mädler, Johann Heinrich: 46
vulcanism: 33-35; terrestrial 73

WXYZ
Wan Hu: 53
weather: *see* climate
weights and measures: 89-93, *90-92, 93*
what to take: 88-89
when to go: 67
Zond spacecrafts: 54-55

ABOUT THE AUTHOR

Carl Koppeschaar studied astronomy and physics before dedicating himself to science writing. In his pursuit of the shadow of the Moon during total solar eclipses, he traveled to the most exotic places on Earth. The only place left to visit was the Moon, where he researched and wrote the *Moon Handbook*. When not exploring the lunar surface, Carl may be found on Earth, where he lives in the centuries-old city of Haarlem, the Netherlands.

MOON TRAVEL HANDBOOKS: THE IDEAL TRAVELING COMPANIONS

Moon Travel Handbooks provide focused, comprehensive coverage of distinct destinations all over the world. Our goal is to give travelers all the background and practical information they'll need for an extraordinary, unexpected travel experience.

Every Handbook begins with an in-depth essay about the land, the people, their history, art, politics, and social concerns—an entire bookcase of cultural insight and introductory information in one portable volume. We also provide accurate, up-to-date coverage of all the practicalities: language, currency, transportation, accommodations, food, and entertainment. And Moon's maps are legendary, covering not only cities and highways, but parks and trails that are often difficult to find in other sources.

Moon Travel Handbooks cover North America and Hawaii, Mexico, Central America and the Caribbean, and Asia and the Pacific. To purchase Moon Travel Handbooks, please check your local bookstore or order by phone: (800) 345-5473 Monday-Friday 8 a.m.-5 p.m. PST.

Praise from press around the world for Moon Travel Handbooks:

> "Moon guides are wittily written and warmly personal; what's more, they present a vivid, often raw vision of Asia without promotional overtones. They also touch on such topics as official corruption and racism, none of which rate a mention in the bone-dry, air-brushed, dry-cleaned version of Asia written up in the big U.S. guidebooks."
> —*Far Eastern Economic Review*

> "The finest are written with such care and insight they deserve listing as literature."
> —*American Geographical Society*

> "Moon's greatest achievements may be the individual state books they offer.... Moon not only digs up little-discovered attractions, but also offers thumbnail sketches of the culture and state politics of regions that rarely make national headlines."
> —*The Millennium Whole Earth Catalog*

TRAVEL MATTERS

Travel Matters is Moon Publications' free quarterly newsletter, loaded with specially commissioned travel articles and essays that tell it like it is. Recent issues have been devoted to Asia, Mexico, and North America, and every issue includes:

Feature Stories: Travel writing unlike what you'll find in your local newspaper. Andrew Coe on Mexican professional wrestling, Michael Buckley on the craze for wartime souvenirs in Vietnam, Kim Weir on the Nixon Museum in Yorba Linda.

Transportation: Tips on how to get around. Rick Steves on a new type of Eurail pass, Victor Chan on hiking in Tibet, Joe Cummings on how to be a Baja road warrior.

Health Matters: Articles on the most recent findings by Dr. Dirk Schroeder, author of *Staying Healthy in Asia, Africa, and Latin America*. Japanese encephalitis, malaria, the southwest U.S. "mystery disease" . . . forewarned is forearmed.

Book Reviews: Informed assessments of the latest travel titles and series. The Rough Guide to *World Music,* Let's Go vs. Berkeley, Dorling Kindersley vs. Knopf.

The Internet: News from the cutting edge. The Great Burma Debate in rec.travel.asia, hotlists of the best WWW sites, updates on Moon's massive "Road Trip USA" exhibit.

There are also booklists, Letters to the Editor, and anything else we can find to interest our readers, as well as Moon's latest titles and ordering information for other travel products, including Periplus Travel Maps to Southeast Asia.

To receive a free subscription to *Travel Matters,* call (800) 345-5473, write to Moon Publications, P.O. Box 3040, Chico, CA 95927-3040, or e-mail travel@moon.com.

Please note: Subscribers who live outside the United States will be charged $7.00 per year for shipping and handling.

MOON TRAVEL HANDBOOKS

ASIA AND THE PACIFIC

Bali Handbook (3379)	$12.95
Bangkok Handbook (0595)	$13.95
Fiji Islands Handbook (0382)	$13.95
Hong Kong Handbook (0560)	$15.95
Indonesia Handbook (0625)	$25.00
Japan Handbook (3700)	$22.50
Micronesia Handbook (3808)	$11.95
Nepal Handbook (3646)	$12.95
New Zealand Handbook (3883)	$18.95
Outback Australia Handbook (3794)	$15.95
Philippines Handbook (0048)	$17.95
Southeast Asia Handbook (0021)	$21.95
South Pacific Handbook (3999)	$19.95
Tahiti-Polynesia Handbook (0374)	$13.95
Thailand Handbook (3824)	$16.95
Tibet Handbook (3905)	$30.00
*Vietnam, Cambodia & Laos Handbook (0293)	$18.95

NORTH AMERICA AND HAWAII

Alaska-Yukon Handbook (0161)	$14.95
Alberta and the Northwest Territories Handbook (0676)	$17.95
Arizona Traveler's Handbook (0536)	$16.95
Atlantic Canada Handbook (0072)	$17.95
Big Island of Hawaii Handbook (0064)	$13.95
British Columbia Handbook (0145)	$15.95
Catalina Island Handbook (3751)	$10.95
Colorado Handbook (0137)	$17.95
Georgia Handbook (0609)	$16.95
Hawaii Handbook (0005)	$19.95
Honolulu-Waikiki Handbook (0587)	$14.95
Idaho Handbook (0617)	$14.95
Kauai Handbook (0013)	$13.95
Maui Handbook (0579)	$14.95
Montana Handbook (0544)	$15.95
Nevada Handbook (0641)	$16.95
New Mexico Handbook (0153)	$14.95
Northern California Handbook (3840)	$19.95

Oregon Handbook (0102). $16.95
Texas Handbook (0633). $17.95
Utah Handbook (0684) $16.95
Washington Handbook (0552). $15.95
Wyoming Handbook (3980) $14.95

MEXICO
Baja Handbook (0528). $15.95
Cabo Handbook (0285) $14.95
Cancún Handbook (0501). $13.95
Central Mexico Handbook (0234) $15.95
*Mexico Handbook (0315) $21.95
Northern Mexico Handbook (0226) $16.95
Pacific Mexico Handbook (0323) $16.95
Puerto Vallarta Handbook (0250) $14.95
Yucatán Peninsula Handbook (0242). $15.95

CENTRAL AMERICA AND THE CARIBBEAN
Belize Handbook (0370). $15.95
Caribbean Handbook (0277) $16.95
Costa Rica Handbook (0358). $18.95
Jamaica Handbook (0129) $14.95

INTERNATIONAL
Egypt Handbook (3891). $18.95
Moon Handbook (0668) $10.00
Moscow-St. Petersburg Handbook (3913). $13.95
Staying Healthy in Asia, Africa, and Latin America (0269) . . $11.95

* New title, please call for availability

PERIPLUS TRAVEL MAPS
All maps $7.95 each

Bali	Hong Kong	Singapore
Bandung/W. Java	Java	Vietnam
Bangkok/C. Thailand	Ko Samui/S. Thailand	Yogyakarta/C. Java
Batam/Bintan	Penang	
Cambodia	Phuket/S. Thailand	

WHERE TO BUY MOON TRAVEL HANDBOOKS

BOOKSTORES AND LIBRARIES: Moon Travel Handbooks are sold worldwide. Please write to our sales manager for a list of wholesalers and distributors in your area.

TRAVELERS: We would like to have Moon Travel Handbooks available throughout the world. Please ask your bookstore to write or call us for ordering information. If your bookstore will not order our guides for you, please contact us for a free title listing.

> Moon Publications, Inc.
> P.O. Box 3040
> Chico, CA 95927-3040 U.S.A.
> Tel: (800) 345-5473
> Fax: (916) 345-6751
> E-mail: travel@moon.com

IMPORTANT ORDERING INFORMATION

PRICES: All prices are subject to change. We always ship the most current edition. We will let you know if there is a price increase on the book you order.

SHIPPING AND HANDLING OPTIONS: Domestic UPS or USPS first class (allow 10 working days for delivery): $3.50 for the first item, 50 cents for each additional item.

EXCEPTIONS:

Tibet Handbook and *Indonesia Handbook* shipping $4.50; $1.00 for each additional *Tibet Handbook* or *Indonesia Handbook*.

Moonbelt shipping is $1.50 for one, 50 cents for each additional belt.

Add $2.00 for same-day handling.

UPS 2nd Day Air or Printed Airmail requires a special quote.

International Surface Bookrate 8-12 weeks delivery: $3.00 for the first item, $1.00 for each additional item. Note: Moon Publications cannot guarantee international surface bookrate shipping. Moon recommends sending international orders via air mail, which requires a special quote.

FOREIGN ORDERS: Orders that originate outside the U.S.A. must be paid for with either an international money order or a check in U.S. currency drawn on a major U.S. bank based in the U.S.A.

TELEPHONE ORDERS: We accept Visa or MasterCard payments. Minimum order is US$15.00. Call in your order: (800) 345-5473, 8 a.m.-5 p.m. Pacific Standard Time.

ORDER FORM

Be sure to call (800) 345-5473 for current prices and editions or for the name of the bookstore nearest you that carries Moon Travel Handbooks • 8 a.m.–5 p.m. PST.
(See important ordering information on preceding page.)

Name: _____ Date: _____

Street: _____

City: _____ Daytime Phone: _____

State or Country: _____ Zip Code: _____

QUANTITY	TITLE	PRICE

Taxable Total_____
Sales Tax (7.25%) for California Residents_____
Shipping & Handling_____
TOTAL_____

Ship: ☐ UPS (no P.O. Boxes)   ☐ 1st class   ☐ International surface mail
Ship to: ☐ address above   ☐ other _____

_____

Make checks payable to: **MOON PUBLICATIONS, INC**. P.O. Box 3040, Chico, CA 95927-3040 U.S.A. We accept Visa and MasterCard. **To Order**: Call in your Visa or MasterCard number, or send a written order with your Visa or MasterCard number and expiration date clearly written.

Card Number: ☐ **Visa**   ☐ **MasterCard**

☐ ☐ ☐ ☐   ☐ ☐ ☐ ☐   ☐ ☐ ☐ ☐   ☐ ☐ ☐ ☐

Exact Name on Card: _____

Expiration date:_____

Signature: _____

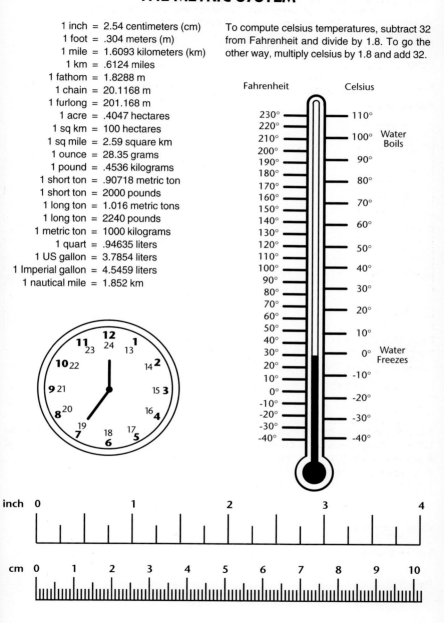